JOURNAL OF
GREEN ENGINEERING

Volume 5, No. 2 (April 2015)

Published, sold and distributed by:
River Publishers
Niels Jernes Vej 10
9220 Aalborg Ø
Denmark

River Publishers
Lange Geer 44
2611 PW Delft
The Netherlands

Tel.: +45369953197
www.riverpublishers.com

Journal of Green Engineering is published four times a year.
Publication programme, 2015: Volume 5 (4 issues)

ISSN 1904-4720 (Print Version)
ISSN 2245-4586 (Online Version)
ISBN 978-87-93379-72-5

JOURNAL OF GREEN ENGINEERING

Volume 5 No. 2 April 2015

Quality of Service Aware Multi-Hop Relay Networks for Green Radio Communication

M. Arthi*, P. Arulmozhivarman and K. Vinoth Babu

School of Electronics Engineering, VIT University, Vellore, Tamil Nadu, India
*Corresponding Author: arthimdas@gmail.com

Received 24 June 2015; Accepted 29 September 2015;
Publication 15 October 2015

Abstract

In recent years, issues on greenhouse pollution and power consumption related to the operation of information communication technology (ICT) devices have triggered a huge amount of research work towards green radio communication. Many of the power efficient solutions will introduce cost performance trade-offs. The energy efficient solution which also takes care of the quality requests of the mobile customer is really required for the 5G based green radio communication. It is identified that enhanced nodeBs (eNB) are power inefficient and not suited for many deployment scenarios. Multi-hop relay (MHR) network is the suitable alternate for eNB especially in the coverage holes and cell edges. But MHR network suffers by various quality issues like site planning and path selection. Many of the site planning and path selection schemes will fail under network imbalance conditions. In this work, we have proposed a hybrid throughput oriented (TO) – load aware spectral efficient routing (LASER) scheme which offers high performance even under network load imbalance conditions. The simulation results proved that the proposed scheme is more suited for future green radio communication.

Keywords: Green radio, LASER, path selection, site planning, TO scheme.

Journal of Green Engineering, Vol. 5, 85–106.
doi: 10.13052/jge1904-4720.521

1 Introduction

In recent years, green radio communication has become one of the hot research topics among the network operators, manufacturers and researchers [1]. In general, the term green technology is dedicated for the efforts to reduce the emissions of greenhouse gases from the industry. For telecommunication operators, the objectives of green approaches are to gain extra commercial benefits by reducing operating expenses related to energy cost. 5G wireless design is enormously propelled by energy efficiency perspectives. Green communication has pulled in a considerable measure of consideration because of rising electricity cost of general network operations and its unfriendly impact on environment due to CO_2 emissions [2].

It has been noticed that the present wireless networks are not power efficient especially the eNBs [1, 2]. The increase in the number of mobile users and the corresponding data rate demands, will increase the overall power consumption of the ICT industries. It is also noticed that 4 % of the global CO_2 emissions are from ICT industries. The cellular networks alone contribute 12 % of the ICT CO_2 emissions. In a cellular communication networks, almost 80 % of the power is consumed by eNB sites. The power consumption of the eNB depends on the radio power transmitted, power consumed by digital signal processing transceivers, cooling units and the power back up units. The power consumed by the cooling units and back up units is more than 55 % of the power consumed by the eNB. Present days, the number of eNBs has reached more than 4 million. The amount of CO_2 emissions from eNB sites alone reached 78 million tons (approximately equal to the emissions from 15 million cars). It is anticipated that greenhouse gases emission from ICT devices will double or even triple in 2020. Manner et al. proved that in order to offer coverage similar to 2G network, Long term evolution (LTE) has to consume about 60 times amount of energy [3]. The power consumed and the amount of CO_2 emissions by various components of cellular network is displayed in Table 1.

Many researchers are working on green communication to reduce the energy bills for telecomm service providers. In literature, there exist five different approaches to minimize the energy cost [5]. They are

1. Improving energy efficiency of hardware components.
2. Turning off the hardware components selectively.
3. Optimizing the energy efficiency of radio transmission.
4. Adopting renewable energy sources.
5. Planning and deploying small cells.

Table 1 Comparison of power consumption and CO_2 emissions by various network components [4]

Components	Power Consumption	CO_2 Emissions (CO_2/a)
User Equipment (UE)	0.1 W	~1 Mt
eNB	1 kW	~30 Mt
Network control (radio network controller, base station controller)	1 kW	<0.5 Mt
Core and servers (mobile switching centre, AAA server, IP core)	10 kW	~7 Mt

The first approach is energy efficient design of hardware components like power amplifier [6]. In a macro eNB, the input power is approximately 1500 W. The air conditioning, signal processing and AC rectifier circuit takes powers of 200 W, 150 W and 150 W respectively. The remaining 65 % of the input power (i.e. 1000 W) alone given to the power amplifier. Based on the efficiency of the power amplifier much power is dissipated as heat. For example, 12 % efficient power amplifier output will be 120 W. The power amplifier output is given as the input to the antenna through a co-axial feeder. This operation introduces 50 % loss. Thus almost 80 % of the input power is dissipated as heat and the useful output power is only around the range 5 % to 20 %. The power amplifier module with 35 % of power efficiency for a small cell wideband code division multiple access (WCDMA) or LTE eNB costs around $75 which covers a radius less than 2 km [7]. Increasing the power efficiency or coverage will also increase the implementation cost.

It has been identified that in LTE networks, 60 % of power consumption is scaled with the traffic load [8]. As per European Telecommunication standards Institute (ETSI) definition, low load eNB requires 10 % of radio frequency (RF) power. Medium load and busy load eNBs require 30 % and 50 % RF powers respectively. Thus in the second approach, eNB or some of the hardware components are put in the sleep mode during non-peak traffic hours. This may reduce unnecessary power consumptions like air conditioning etc. But this sleep mode approach may negatively impact on quality of service (QoS). It may decrease the capacity.

In the third approach, transmission energy is minimized by using the techniques like multi input multi output (MIMO), cognitive radio, energy efficient scheduling, channel coding etc [5, 9]. But again these approaches may increase the computational complexity. In the fourth approach instead of using the conventional energy resources like hydro carbon which produces greenhouse gases, it is suggested to use renewable resources like hydro, wind and solar powers [10]. The network operators in Bangladesh, Nigeria are already using solar operated eNBs [11, 12]. Due to poor road and unsafe conditions, delivering the traditional energy resources like diesel is not always guaranteed. But again purchasing, replacing and installing new equipment's (including man power, transportation, and disruption of normal operation) may increase the cost.

The fifth approach is to introduce small cell networks like micro cell, pico cell or femto cell instead of macro eNBs [13]. The deployment cost and power consumption of small cells are much lower when compared to macro eNBs. The transmission power and deployment cost requirements of various LTE based eNBs are listed in Table 2. These small cells bring additional radio interferences when compared to the conventional homogeneous macro eNBs. Deploying more number of small cells may reverse the trend of energy saving. It also requires more overhead during the transmission.

The green radio requirements should also satisfy the customer demands like increase in average throughput per user and decrease in cost per bit [14]. The additive white Gaussian noise (AWGN), interference and shadowing effects will degrade the received signal quality and coverage [15]. The degradations can be overcome by deploying more number of eNB in the geographic area. Increase in the number of eNBs in a given geographic area, will increase the network cost, power consumption and interference. Thus to ensure green radio communication, the deployment of eNB should be minimized as much as possible without sacrificing the QoS.

MHR network is one of the suitable solutions to achieve energy efficiency. Long term evolution-advanced (LTE-A) Rel-10 and IEEE 802.16j standards have introduced the concept of MHR networks, where the Relay Nodes (RN) are used to improve the network performance along with the eNBs [16]. The transmission power and cost of the RNs are much smaller than eNB. The introduction of RNs may shorten the distance between the communicating nodes. This in turn reduces the transmission power required. RNs deployment is very useful in the areas where the eNB backhaul solutions

Table 2   Comparison of LTE eNBs [13]

eNB	Transmission Power	Cost
Macro	20 W–160 W	397,800 Euros
Micro	2 W–20 W	42,200 Euros
Pico	250 mW–2W	12,400 Euros

(fibers, microwave links etc.) are unavailable and very expensive [17]. RNs can also be deployed in the areas where site acquisition for eNB deployment is very difficult. Since the deployment and removal can be done faster than the eNB, RNs are more preferable for temporary deployments than eNB. Deployment of RNs in the coverage holes will improve the received signal quality of surrounding UE. This in turn increases the total system capacity without much increasing the capital and operational expenses of the network operator. Because of the more beneficial nature and less power consumption, RNs are more likely to be deployed in the geographic area than the eNBs. This turns to be one of the best solutions for future green radio communication. But there exist some major issues like site planning and RN selection in MHR networks [16]. Until these issues addressed, MHR networks may not satisfy the customer demands in terms of throughput and cost. The unsatisfied customer will move to the competing operator.

Site planning and deployment of RNs in appropriate locations may reduce the total power required to serve the UEs [18]. Appropriate site planning will increase the system throughput and coverage. There exist several studies for eNB and RN site planning in literature [18–25]. Y. Yu et al. proposed a cost function based eNB and RN site planning scheme for IEEE 802.16j networks [19]. S-J. Kim et al. have proposed a cost effective coverage extension solution for mobile worldwide interoperability for microwave access (WiMAX) system [20]. The proposed solution analyzes the total deployment cost first and then determines the optimum number of eNBs and RNs based on the traffic demands. D. Yang et al. proposed a RN placement algorithm which minimizes the number of RNs to be deployed, using cooperative communication scheme and satisfies all the data rate requirements of the UEs [21]. H-C. Lu et al. proposed a heuristic scheme which determines the number of RN deployment locations for a given UE distribution and deployment budget [22]. H-C. Lu & W. Liao proposed a joint eNB and RN deployment scheme for IEEE 802.16j networks [23]. The proposed scheme maximizes the system throughput by ensuring the total deployment cost within the limits of maximum allowed deployment budget. This scheme suffers by network load balancing issues. J-Y. Chang and Y-S. Lin proposed a novel uniform clustering based eNB and RN deployment scheme [24]. This scheme works in two phases, namely eNB-RN selection phase and eNB-RN deployment phase. This scheme provides reasonable throughput and coverage by balancing the network load between different eNBs. All the above mentioned schemes are computationally complex and many of the schemes encourage more eNB deployment which is not supported in green radio communication. L-C. Wang et al. proposed a low complex RN deployment scheme for IEEE 802.16j networks to maximize the network throughput [25]. The authors used TO and signal strength oriented (SSO) RN deployment schemes in their work. The results showed that the TO schemes offer more throughput than the SSO schemes. Thus in our proposed work, we adopt TO based RN deployment scheme to maximize the network throughput.

Once RNs are deployed, the performance improvements will not only depend on the site planning [21]. There exist a trade-off between system throughput and fairness. To offer high quality services to every user, appropriate set of RNs selection is important. There exist two different scenarios of MHR networks where RN selection is important. In the first scenario, an UE may be in the coverage area of both eNB and RN. In the second scenario, an UE may be covered by multiple RNs. But each UE should get the service from only one node. Thus, the path selection is an important issue in MHR networks. Improper path selection may lead to loss in system throughput and increase in the signal transmission delay [26, 27]. In literature, there exist many path selection schemes. Many of the powerful schemes like radio resource utilization index (RRUI), are also not addressed the link overloading issue. S-S. Wang et al. proposed a LASER path selection scheme which uses link cost as a metric for path selection [28]. This scheme offers better performance under link overloading conditions. Thus, in the proposed work, we adopt LASER based path selection scheme to address the link overloading issue.

The rest of the script is organized in the following order. The system model and the novel TO-LASER schemes are described in Sections 2 and 3 respectively. The simulation results are discussed in Section 4 and the paper is concluded in Section 5.

2 System Model

MHR network consists of three major nodes namely eNB, RN and UE [29]. LTE-A and IEEE 802.16j standards allow two different RN deployment scenarios [16]. In the first scenario, the RNs are deployed within the coverage area of eNB. These RNs are transparent RNs. This type of deployment will solve the coverage problems of the UEs especially in the coverage holes like indoor, underground etc. In the second scenario, RNs are deployed near the cell edge to extend the coverage. Such RNs are called non transparent RNs. In this work, site planning and RNs selection is done for both transparent and non-transparent RNs. All eNBs are connected to the mobile switching centre

(MSC) which is acting as the gateway between wired and wireless networks. A typical MHR network is shown in Figure 1.

Based on the mobility, there exist three different types of RNs namely fixed RN (FRN), Nomadic RN (NRN) and Mobile RN (MRN). When there is a slow and group mobility, NRNs and MRNs are preferred respectively. NRNs are deployed in slow moving vehicles. MRNs are deployed in systems where there exist medium to high mobility like public transportation systems. Since the mobility of NRN is less, it can support certain number of users for certain amount of time without requiring more number of handoffs. Similarly, the group of users travelling in a train or bus will be getting the service from the MRN deployed in that transportation system. The connected users will be getting the service from the same MRN for most of the time. But the proposed work is limited to FRN. Identifying different types of RNs for different sites is out of scope for this work. The site planning for RNs should guarantee two features. The first one is, there should be a line of sight (LOS) from all RNs to eNB and the second one is there should be a non-line of sight (NLOS) from a RN to other interfering RNs. It is also assumed that eNB is deployed in the optimum location. Using the TO based site planning rule, the FRNs should be placed in the optimum distance from the eNB where the system throughput

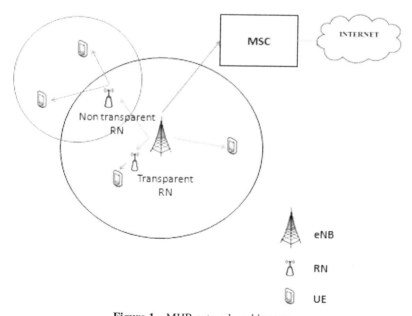

Figure 1   MHR network architecture.

is maximum. At the end, the objective is to maximize the coverage ratio i.e each UE within the given geographic area must be connected to a eNB or RN or both.

For simple analysis, a free space propagation model is used to measure the received signal to noise ratio (SNR). The received SNR is given by [24],

$$SNR(dB) = 10 \log_{10} \left(\frac{P_R}{\sigma_N^2} \right) \qquad (1)$$

where σ_N^2 is the noise variance and P_R is the received signal power which is given by

$$P_R = P_T \left(\frac{c}{4\pi f_c d} \right)^2 \qquad (2)$$

where P_T is the transmitted signal power, f_c is the carrier frequency, c is the velocity of light and d is the distance between any two communicating nodes. The distance, received SNR and the corresponding burst profile is listed in Table 3.

The two phase transmission of the MHR network may reduce the system capacity [25]. It may also lead to unnecessary delay in transmission which is not desired in time bounded applications. To maximize the capacity and to minimize the transmission delay, the eNB has to decide for an indirect transmission, only when it is needed. This problem is more severe when the UEs are connected by both eNB and RN or by multiple RNs. Based on the TO rule, the indirect transmission is preferred as long as,

$$R_{eNB-RN-UE} > R_{eNB-UE} \qquad (3)$$

Table 3   Modulation and coding scheme (MCS) for link adaptation [24]

Burst Profile	Coding Rate	Modulation	Distance between the Nodes (km)	SNR (dB)	Data Rate (Mbps)
1	1/2	BPSK	3.20	3.00	1.269
2	1/2	QPSK	2.70	6.00	2.538
3	3/4	QPSK	2.50	8.50	3.816
4	1/2	16-QAM	1.90	11.50	5.085
5	3/4	16-QAM	1.70	15.00	7.623
6	2/3	64-QAM	1.30	19.00	10.161
7	3/4	64-QAM	1.20	21.00	11.439

where $R_{eNB-RN-UE}$ is the data transmission rate of indirect transmission and R_{eNB-UE} is the direct data transmission rate.

The data transmission rate of indirect transmission is given by,

$$R_{eNB-RN-UE} = \frac{D}{t_{eNB-RN-UE}} = \frac{D}{\frac{D}{R_{eNB-RN}} + \frac{D}{R_{RN-UE}}}$$

$$= \left(\frac{1}{R_{eNB-RN}} + \frac{1}{R_{RN-UE}} \right)^{-1} \tag{4}$$

where D is the packet size and $t_{eNB-RN-UE}$ is the time taken for indirect transmission. Based on the TO scheme, the data transmission rate for an UE is

$$R = Max\left(R_{eNB-RN-UE}, R_{eNB-UE} \right) \tag{5}$$

The average system capacity can be obtained using,

$$C = \frac{\sum\limits_{k=1}^{M} R(k)}{M} \tag{6}$$

where M is the total number of UEs in the given geographic area and $R(k)$ is the data transmission rate of the kth UE.

The concept of TO selection rule is illustrated in Figure 2. Based on (5), there are chances that multiple UEs may choose a single RN for service. The functional and capacity limitations of RNs may introduce link overloading issue [28]. This will reduce the net system throughput when it is

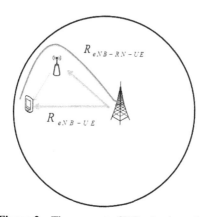

Figure 2   The concept of TO selection rule.

left uncared. The TO selection rule will not address the link overloading issue. A LASER based RN selections rule will address the link overloading issue when compared to other RN selection rules. The concept of link overloading in MHR network is shown in Figure 3.

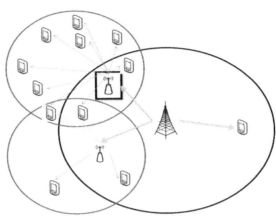

New services

Link overloaded RN

Figure 3   Link overloading.

Table 4   Notations list

Notations	Descriptions
$C(i)$	network capacity at i^{th} distance
C_{RN}	capacity of the RN
X	coordinates of eNB
Y_k	coordinates of the kth UE
Z_j^i	coordinates of jth RN under i^{th} distance
d_i	set of ith possible distance for RNs deployment
d_{max}	maximum possible eNB-RN separation distance
d_{opt}	optimum distance for RNs deployment with maximum network capacity
$e1$	distance between eNB to j^{th} RN for i^{th} distance
$e2$	distance between j^{th} RN and k^{th} UE
$e3$	distance between eNB and k^{th} UE
l	number of possible eNB RN distances
m	number of RNs in the geographic area
z_j	j^{th} RN

3 The Proposed TO-LASER Scheme

In the proposed scheme, along the vertices of eNB, the set of possible equal distances is taken for RNs deployment. The algorithm initially calculates the distance between eNB and RN. It also calculates the distance between each UE with all the RNs in the geographic area. Based on this, each UE identifies the nearby RN. The direct path distance i.e the distance between eNB and UE

TO-LASER Algorithm:
Input: X, Y_k, $d = \{d1, ... d_{\max}\}$, M, m, C_{RN}
Output: d_{opt}
1. for $i = 1$ *to* l
2. $e1 = \|X - d_i\|$
3. for $k = 1$ *to* M
4. for $j = 1$ *to* m
5. $e2 = \|Z_j^i - Y_k\|$
6. find j corresponding to min $e2$
7. $e3 = \|X - Y_k\|$
8. calculate $R(k)$ using (5)
9. if $R(k) = R_{eNB-UE}$
10. $eNB \leftarrow UE(k)$
11. elseif $R(k) = R_{eNB-RN-UE}$
12. if $count(z_j) < C_{RN}$
13. $z_j \leftarrow UE(k)$
14. Else
15. Find j corresponding to the next min $e2$ such that $UE(k) \in z_j$
16. End
17. Iterate steps 8 to 16, until each UE is assigned to either eNB or RN
18. End
19. End
20. End
21. Using (6) calculate C.
22. End
23. $d_{opt} = index\ [\max(C)]$ (7)

is also calculated. Based on (5), the necessity for indirect communication is tested. To check the link overloading condition, the capacity of the nearby RN is checked. If it has less number of UEs compared to its capacity, then the particular UE is assigned to the corresponding RN. Otherwise the next nearby RN is checked for its capacity. This process is repeated, until each UE identifies a RN without link overloading. After assigning all the UEs to either eNB or to any of the nearby RNs, the overall network capacity can be obtained using (6). This process is repeated for all possible set of distances taken for the RNs deployment. Finally the distance for which the maximum network capacity is attained is assumed to be the optimum location for RNs deployment under link overloading condition.

4 Simulation Results and Discussion

In this section, a comparison of different path selection schemes in terms of fairness, spectral efficiency and average transmission power per UE is done. The schemes considered are no relaying, RRUI relaying, LASER relaying and joint TO-LASER relaying. The simulations are repeated for 100 different scenarios and the average results are displayed in this work. The following parameters and assumptions are considered for the simulations:

- The geographic region is assumed to be of 10 km × 10 km.
- The MHR network consists of three major nodes such as eNB, RN and UE.
- RN is considered to be FRN.
- It is assumed that eNB is located in an optimum location.
- eNB and RNs use Omnidirectional antennas.
- eNB and RNs are assumed to have a coverage radius of 3.2 km and 1.9 km respectively.
- 100 UEs are deployed randomly in the geographic region with non-uniform traffic demands.
- It is assumed that all the RNs are in the LOS with eNB.
- The RNs are placed with sufficiently far distance, so that the co-channel interference among them is minimum.
- The candidate locations of the RNs are assumed to be on the direction of the vertices of the hexagonal eNB.
- The path loss calculations are done based on the free space propagation model.

- It is assumed that perfect channel state information (CSI) is available at the transmitter and receiver.
- No particular mobility model is considered in this work.

The average system throughput (Mbps) vs. eNB-RN separation distance is showed in Figure 4. The proposed TO-LASER scheme is executed to find the optimal distance for RNs deployment. Since there exist only direct communication in no relaying scheme, the average system throughput is constant irrespective of the eNB-RN separation distance. For the TO-LASER scheme, the average system throughput increases with the increase in eNB-RN separation distance up to 1.9 km and then it decreases. The system throughput is maximum at 1.9 km. This is almost same for different UE distributions with different traffic demands. Thus the RNs are deployed at the distance of 1.9 km from the eNB. We assume hexagonal cell shape and the RNs are deployed on the directions of the vertices of the hexagonal. This is clear from Figure 5.

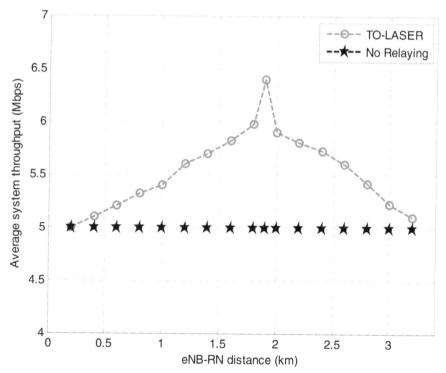

Figure 4 Optimal RN location identification using TO-LASER based selection rule.

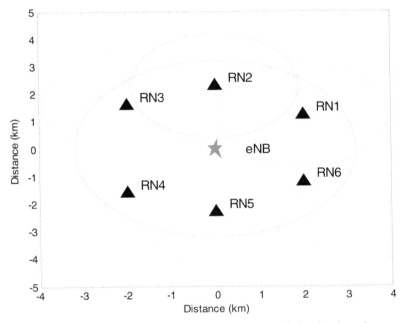

Figure 5   eNB and RNs locations after TO-LASER based site planning scheme.

For comparison, we introduce utilization termed ρ which represents the congestion level of the traffic load in the system. This factor defines the proportion of the system resources used by the traffic in a system.

$$\rho = \frac{\gamma}{\eta} \qquad (8)$$

where γ and η are the mean arrival rate and the mean service rate of a system. The value of utilization in the simulation ranges from 0.1 to 1.0 with a step of 0.1.

Figure 6 shows the average spectral efficiency (b/s/Hz) of various schemes under different utilization. The average spectral efficiencies attained by no relaying, RRUI relaying, LASER relaying and TO-LASER relaying schemes are 0.4758, 0.8405, 1.1324 and 1.2574 respectively. The spectral efficiency performance of LASER relaying is far better than the other two schemes (no relaying and RRUI relaying) under link overloading conditions. The inclusion of TO based site planning along with the LASER based relaying, improves the average spectral efficiency to 9.9729 % than LASER relaying scheme.

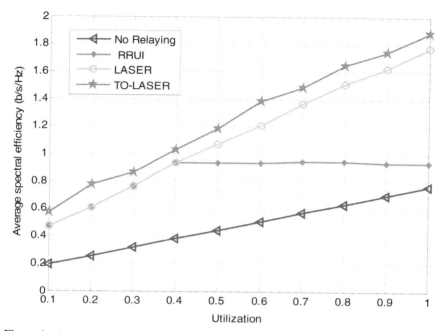

Figure 6 Average spectral efficiency (b/s/Hz) comparison for various path selection schemes.

Figure 7 indicates the fairness of various schemes under different utilization. Fairness determines whether the network under different utilization is receiving a fair share of system resources. It is given by,

$$F = \frac{\left(\sum\limits_{k=1}^{M} R(k)\right)^2}{M \sum\limits_{k=1}^{M} (R(k))^2} \tag{9}$$

It shows that the fairness of TO-LASER and LASER schemes remain same for different utilization values. It shows that all the UEs under the eNB coverage area are receiving a fair share of resources. For RRUI relaying, the fairness begins to drop gradually when it reaches a utilization of 0.4. This is due to the limited link capacity. Even though the no relaying scheme attains maximum fairness under different utilization, it is not preferred due to low spectral efficiency.

The total power consumption is the addition of transmission power and power consumed by the hardware equipment. The downlink data transmission rate of a kth UE is given by,

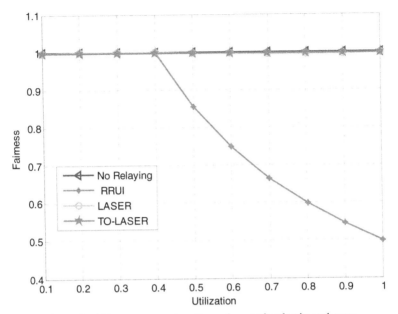

Figure 7   Fairness comparison for various path selection schemes.

$$R_{eNB-UE(k)} = B_{eNB-UE(k)} \log_2 \left(1 + \frac{P_{eNB-UE(k)}}{\sigma_N^2} \left(\frac{c}{4\pi f_c d_{eNB-UE(k)}} \right)^2 \right)$$

(10)

where $B_{eNB-UE(k)}$ is the bandwidth allocated for downlink transmission between eNB and k^{th} UE and $d_{eNB-UE(k)}$ is the distance between eNB and k^{th} UE. The power required for downlink transmission between eNB and k^{th} UE is given by,

$$P_{eNB-UE(k)} = \left(2^{\frac{R_{eNB-UE(k)}}{B_{eNB-UE(k)}}} - 1 \right) \left(\frac{4\pi f_c d_{eNB-UE(k)}}{c} \right)^2 \sigma_N^2$$

(11)

The introduction of RNs will lessen the transmission distance between the communicating nodes. As per (11), the transmission power is directly proportional to the distance. Thus the properly deployed RNs will reduce the total transmission power for achieving a desired fixed rate of transmission.

The total transmission power for no relay scheme is calculated using,

$$P_T = \sum_{k=1}^{M} P_{eNB-UE(k)} \tag{12}$$

The total transmission power for MHR is calculated using,

$$P_T = \sum_{k_1=1}^{M_1} P_{eNB-UE(k_1)} + \sum_{k_2=1}^{M_2} P_{eNB-RN-UE(k_2)} \tag{13}$$

where M_1 and M_2 are the number of UEs connected directly and indirectly with eNB. $P_{eNB-RN-UE(k_2)}$ is the transmission power required for an indirect transmission.

For 100 different UE distributions with utilization 0.5, the average transmission power per UE (in W) is shown in Figure 8. The graph is developed based on the assumption of average AWGN with power spectral density of -174 dBm/Hz and 100 UEs. The average transmission power per UE for no relaying, RRUI, LASER, TO-LASER schemes are -46.95 dBm, -50.41 dBm, -51.22 dBm, -53.49 dBm respectively. We need a minimum of four eNBs to cover the considered geographic area. The total amount of power required by

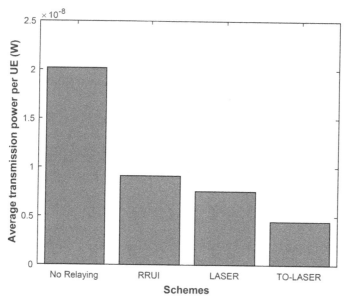

Figure 8 Average transmission power per UE (W) comparison for different schemes.

the eNBs is approximately 6000 W. When we go for TO-LASER scheme, we need one eNB and six RNs which consume the total power approximately of 2500 W. This power consumption is pretty much same for other relay based schemes considered. Be that as it may, the other relay based schemes, require more transmission power than the TO-LASER scheme. Thus the total power required for TO-LASER is much lesser than the other schemes.

5 Conclusions

This paper has discussed the importance of RN based MHR networks for green radio communications. There exist some severe issues like site planning and path selection in MHR networks. In this work, a hybrid TO-LASER site planning scheme is proposed which offers high performance than the conventional LASER based scheme even at link overloading conditions. It shows a 9.9729 % improvement in spectral efficiency over the conventional LASER scheme. The proposed scheme also offers better fairness in terms of resource allocation. The average transmission power per UE for TO-LASER scheme is −53.49 dBm which is much lower than the other considered schemes. All the relay based schemes require only 40 % of the total power consumed by no relaying scheme. The proposed deployment and path selection concepts can be extended to femto relays. The real time impairments like channel estimation error and channel information feedback delay are not considered in this work. The path selection scheme used here is for downlink. The promising results pave path for further investigations on uplink path selection schemes under various channel impairments and co channel interference conditions.

References

[1] P. Serrano; A. de la Oliva; P. Patras; V. Mancuso; A. Banchs, Greening Wireless Communications: Status and Future Directions, *Comput. Commun.*, **35**:1651–1661 (2012).

[2] I. Ahmed; M. M. Butt; C. Psomas; A. Mohamed; I. Krikidis; M. Guizani, Survey on Energy Harvesting Wireless Communications: Challenges and Opportunities For Radio Resource Allocation. *Comput. Netw.*, **88**:234–248 (2015).

[3] J. Manner; M. Luoma; J. Ott; J. Hamalainen, Mobile networks unplugged. In *Proceedings of the 1st International Conference on Energy-Efficient Computing and Networking*, pp. 71–74 (2010).

[4] A. T. Clark, Nokia Siemens Networks Environmentally Sustainable Business Good green business sense [Online]. Available at http://www.istemobility.org/WorkingGroups/Applications/Meeetings-Events/2010-05-19_Athens/AClark.pdf.

[5] J. Wu; Y. Zhang; M. Zukerman; E. Yung, Energy-efficient Base-stations Sleep-mode Techniques in Green Cellular Networks: A Survey, *IEEE Commun. Surv. Tuts.*, **17**(2):803–826 (2015).

[6] C. Han et al., Green Radio: Radio Techniques to Enable Energy-efficient Wireless Networks. *IEEE Commun. Mag.*, **49**(6):46–54 (2011).

[7] TriQuint, 'RF Power Amplifier Module: TGA2450-SM' [Online] Available at http://store.triquint.com/ProductDetail/TGA2450-SMTriQuint-Semiconductor-Inc/472292/.

[8] I. Ashraf; F. Boccardi; L. Ho, Sleep Mode Techniques for Small Cell Deployments. *IEEE Commun. Mag.*, **49**(8):72–79 (2011).

[9] D. Feng et al., A Survey of Energy-efficient Wireless Communications. *IEEE Commun. Surv. Tuts.*, **15**(1):167–178 (2013).

[10] Y.-K. Chia; S. Sun; R. Zhang, Energy Cooperation in Cellular Networks with Renewable Powered Base Stations, *IEEE Trans. Wireless Commun.* **13**(12):6996–7010 (2014).

[11] Y. Chan, China's Huawei to Supply Solar-Powered Base Stations to Bangladesh. Business Green, London, UK [Online]. Available at http://www.businessgreen.com/bg/news/1802444/, chinas-huawei-supply-solar-powered-base-stations-bangladesh (2009).

[12] C. Okoye. Airtel Base Stations to be Solar Powered, Daily Times Nigeria, Lagos, Nigeria [Online]. Available at http://dailytimes.com.ng/article/airtel-base-stations-be-solar-powered (2011).

[13] Y.-C. Wang; C.-A. Chuang, Efficient eNB Deployment Strategy for Heterogeneous Cells in 4G LTE Systems. *Comput. Netw.* **79**: 297–312 (2015).

[14] 3GPP TR 36.806 V9.0.0 3rd Generation Partnership Project, Technical Specification Group Radio Access Network, Evolved Universal Terrestrial Radio Access (E-UTRA), Relay architectures for E-UTRA (LTE-Advanced) (Release 9), http://www.qtc.jp/3GPP/Specs/36806-900.pdf, 2010-13.

[15] M. Arthi; P. Arulmozhivarman; K. VinothBabu; J. Jose Joy; E. Mariam George, Technical Challenges in Mobile Multi-hop Relay Networks. *Int. J. Appl. Eng. Res.*, **10**:26025–26036 (2015).

[16] M. Arthi; P. Arulmozhivarman; K. VinothBabu; G. Ramachandra Reddy; D. Barath, Techniques to Enhance the Quality of Service of Multi hop Relay Networks. *Proc. Comput. Sci.*, **46**:973–980 (2015).

[17] I. F. Akyildiz; E. Chavarria-Reyes; D. M. Gutierrez-Estevez; R. Balakrishnan, LTE-Advanced and the Evolution to Beyond 4G (B4G) systems. *Phys. Commun.*, **10**:31–60 (2014).

[18] B.-J. Chang; Y.-H. Liang; S.-S. Su, Analyses of Relay Nodes Deployment in 4g Wireless Mobile Multihop Relay Networks, *Wireless PersCommun.* doi: 10.1007/s11277-015-2443-x (2015).

[19] Y. Yu; S. Murphy; L. Murphy, Planning base station and relay station locations in IEEE 802.16j multi-hop relay networks. In *Consumer communications and networking conference*, pp. 922–926 (2008).

[20] S.-J. Kim; S.-Y. Kim; B.-B. Lee; S.-W. Ryu; H.-W. Lee; C.-H. Cho, Multi-hop Relay Based Coverage Extension in the IEEE802.16j Based Mobile WiMAX Systems. In *Fourth international conference on networked computing and advanced information management*, Vol. 1, pp. 516–522 (2008).

[21] D. Yang; X. Fang; G. Xue; J. Tang, Relay Station Placement for Cooperative Communications in WiMAX Networks. In *Proceedings of the IEEE global telecommunications conference, (GLOBECOM)* (2010).

[22] H.-C. Lu; W. Liao; F. Y-S. Liu, Relay Station Placement Strategy in IEEE 802.16j WiMAX Networks. *IEEE Trans. Commun.*, **59**:151–158 (2011).

[23] H.-C. Lu; W. Liao, Joint Base Station and Relay Station Placement for IEEE 802.16j Networks. In *Proceedings of IEEE global telecommunications conference, (GLOBECOM)* (2009).

[24] J.-Y. Chang; Y.-S. Lin, A Clustering Deployment Scheme for Base Stations and Relay Stations in Multi-Hop Relay Networks. *Comput. Electric. Eng.* **40**(2):407–420 (2014).

[25] L.-C. Wang; W.-S. Su; J.-H. Huang; A. Chen C.-J. Chang, Optimal Relay Location in Multi-Hop Cellular Systems. In *Proceedings of the IEEE wireless communications and networking conference (WCNC)* (2008).

[26] S.-S. Wang; H.-C. Yin; Y.-H. Tsai; S.-T.Sheu, An Effective Path Selection Metric for IEEE 802.16-based Multi-hop Relay Networks. In *Proceedings of the IEEE symposium on computers and communications*, pp. 1051–1056 (2007).

[27] D. D. Couto; D. Aguayo; J. Bicket; R. Morris, A High-throughput Path Metric for Multi-hop Wireless Routing. In *Proceedings of the ACM international conference on mobile computing and networking (MOBICOM)*, pp. 134–146 (2003).

[28] S.-S. Wang, C.-Y. Lien, W.-H. Liao and K.-P. Shih. LASER, A Load-aware Spectral-efficient Routing metric for Path Selection in IEEE 802.16j Multi-hop Relay Networks. Comput. Electric. Eng., **38**:953–962 (2012).

[29] D. Satishkumar; N. Nagarajan, Relay Technologies and Technical Issues in IEEE 802.16j Mobile Multi-hop Relay (MMR) Networks. *J. Netw. Comput. Appl.*, **36**:91–102 (2013).

Biographies

M. Arthi is currently working as a Research Associate in School of Electronics Engineering, VIT University, Vellore, Tamil Nadu, India. She has completed her B.E and M.E degrees from Anna University, Chennai, India in 2010 and 2012 respectively. Her area of interests includes Wireless networks and computer communication networks.

P. Arulmozhivarman is currently working as a Professor in School of Electronics Engineering, VIT University, Vellore, Tamil Nadu, India. He is the program chair for Electronics and Communication Engineering department, Division chair for Digital Signal Processing division and Assistant Dean for School of Electronics Engineering. He has completed his Ph.D degree from National Institute of Technology, Trichy in 2005. His area of interests includes Digital Signal Processing, Digital Image processing, neural networks and pattern recognition etc. He served as one of the editor in several international conference proceedings and he is a reviewer of International Journal of Computer Engineering Research and IEEE Photonics Technology letters. Dr. P. Arulmozhivarman is a member of IEEE signal processing society.

K. Vinoth Babu is currently working as an Associate Professor in School of Electronics Engineering, VIT University, Vellore, Tamil Nadu, India. He has completed his Ph.D and M.Tech degrees from VIT University, Vellore, India in 2014 and 2009 respectively and B.E (ECE) Degree from Anna University, Chennai, India in 2005. His area of interests includes Wireless Digital Communications and Signal Processing.

Measurement of Power Radiation in Base Transceiver Station Using Quad Phone and Quadcopter

Prem Kumar N[1], Raj Kumar A[1], Sundra Anand[1], E. N. Ganesh[2]
and V. Prithiviraj[3]

[1]*Final year B.E. (ECE), Rajalakshmi Institute of Technology, Tamil Nadu, India*
[2]*Dean R&I, Rajalakshmi Institute of Technology, Tamil Nadu, India*
[3]*Professor, Dept. of Electronics and Communication Engg (ECE),
Rajalakshmi Institute of Technology, Tamil Nadu, India*
*Corresponding Authors: {prem0504kumar; raj16993; sundra.ece;
profvpraj}@gmail.com; dean.research@ritchennai.ediu.in*

Received January 2015; Accepted March 2015;
Publication April 2016

Abstract

The communication protocols used in Base Transceiver Station (BTS) could be harmful to human species and other life forms in the ecosystem. The BTS used for these systems could emit radiations beyond safety threshold. Therefore, it is essential to monitor such power radiation levels from time to time. Manual readings at each BTS are strenuous and time consuming work. This paper proposes radiation measurement using Quad Phone assembled within the Quadcopter using android software. Quadcopter is a low cost restricted pay-load machine which suits the measurement of radiation emitted from antenna towers and power lines. An application is developed to monitor the power radiation emitted by each of the bands and the associated communication protocol by utilizing the Quad Phone. It can use CDMA, GSM, 3G and LTE protocols at the designated frequency bands. A Quad Phone incorporating the android application is used to record power radiation levels at several points around a single tower facilitated by the Quadcopter flight navigation system. The Wave Point navigation is a hardware module used to move around a target

Journal of Green Engineering, Vol. 5, 107–128.
doi: 10.13052/jge1904-4720.522

point up to 16 locations and circle back to the initial position. The collected data regarding the radiated signal strength for different protocols are either transmitted through wireless or stored within the Quad Phone which could be retrieved later. The power readings around each base station can be recorded using the Quad Phone and the Quadcopter in the shortest time possible. Hence the combination of the Quad Phone mounted on the Quadcopter provides an excellent monitoring system for auditing the Electromagnetic Radiation (ER) and subsequently determine the Electromagnetic Pollution Index (EPI) from the delineated pockets of pollution regions.

Keywords: Base Transceiver Station (BTS), Code Division Multiple Access (CDMA), Global system for Mobile Communications (GSM), Wave point navigation, Global Positioning System (GPS), Electromagnetic Radiation (ER).

1 Introduction

Quadcopter is a multirotor helicopter and is propelled by four rotors. In order to possess less Kinetic Energy during the flight period, each rotor which is mounted on the Quadcopter should have a smaller diameter when compared to the helicopter rotor. As the Quadcopter has its propeller pitch angle not being subjected to any variation, it results in a simpler design implementation for the copter. At present there are fewer vehicles capable of utilising vertical takeoffs and landing mechanisms [1]. The Base Transceiver Stations emits radiations which is being picked up by the cell phones and enables to establish a connection. Now-a-days in order to increase the cell coverage the service providers could radiate large amount of energy which could be harmful to the human species and other flora and fauna. To monitor over these radiation levels the Indian Ministry of Telecommunication formed a committee called TERM cells where the members of the committee move around each BTS measuring the signal strength for each service providers for every quarter. This method is found to be tedious as it requires large amount of human resource, vehicle field strength measurement and is a time consuming process.

This brought immediate attention and as a solution for this problem it is proposed to develop a frame work with a goal to develop an android application for monitoring the signal strength of the BTS at various locations utilizing a Quad Phone mounted within a Quadcopter. A quad phone is quad core quad band phone that works in multiple radio frequency bands using

GSM/CDMA technology. The need for Quadphone is to measure the signal strength across four different cellular services like GSM, CDMA, HSDPA and LTE. Since Quad Phone is light weight smart phone it can be readily used with a Quadcopter to measure the signal strength and radiation level. The four rotors which are mounted in a cross combination provides the necessary lift force thereby enabling the Quadcopter to be airborne. A major distinguishing factor is due to the controlling mechanism of the Quadcopter which is different from the normal helicopter. Quadcopters possess a significant advantage compared to the fixed wing airplanes due to its maneuverability in all directions, with improved hovering capability and for flying at lower speeds.

The rotors mounted on the quadcopter are distinctly responsible for producing the desired thrust and torque around its center of rotation. A distinguishing point is that its propellers are not at all alike and this enables the quadcopter to operate under two pairs, one for establishing the pushing effect and the second pair for enabling the pulling effect, consequently enabling the resultant net torque to be null, provided all the propellers turn with the same angular velocity, thereby permitting the coptercraft to remain still around its center of gravity [7].

Figures 1 and 2 illustrates the movements of each rotor. Changes in the pitch angle are induced by contrary variation of speeds in propellers 1 and 3

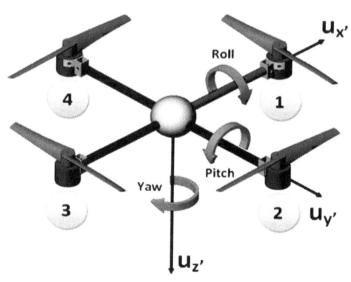

Figure 1 Yaw, pitch and roll rotations of a Quadcopter [7].

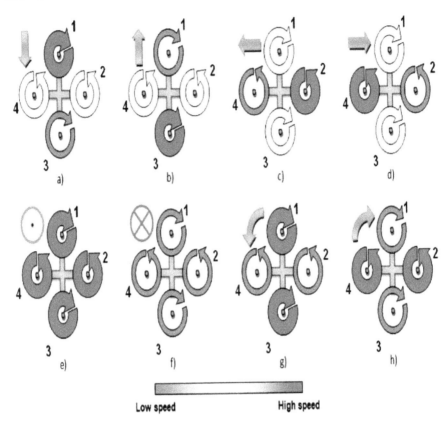

Figure 2   Various movements of a Quadcopter [2].

(see Figure 1), resulting in forward or backwards translation. If we do this same action for propellers 2 and 4, we can produce a change in the roll angle and we will get lateral translation. Yaw is induced by mismatching the balance in aerodynamic torques (i.e. by offsetting the cumulative thrust between the counter-rotating blade pairs). So, by changing these three angles in a Quadcopter we are able to make it maneuver in any direction.

2 Quadcopter Design

The most important target of this particular design process is to arrive at the correct set of requirements for the copter with appropriate payload which could be summarized into a set of specifcations. The specifcations for the Quadcopter prototype are given below:

- Flight autonomy between 15 to 25 minutes.
- On-board controller should have its separate power supply to prevent simultaneous engine and processor failure in case of battery loss.
- Ability to transmit live telemetry data and receive movement orders from a ground station wirelessly, therefore avoiding the use of cables which could become entangled in the aircraft and cause an accident;
- Quadcopter should not fly very far from ground station so there is no need of long range telemetry hardware and the extra power requirements associated with long range transmissions.

Consequently, the main components are:

- 4 electric motors and 4 respective Electronic Speed Controllers (ESC).
- 4 propellers.
- 1 on-board processing unit with inbuilt Accelerometer, Gyroscope, Barometer and Magnetometer.
- 2 on-board power supplies (batteries), one for the motors and another to the processing unit. At this point we will assume that beyond the need to provide electrical power to the motors, we must assure that the brain of the Quadcopter (i.e. the on-board processing unit) remains working well after the battery of the motors has discharged;
- 1 airframe for supporting all the aircraft's components.

2.1 Airframe

The airframe is the mechanical structure of an aircraft that supports all the components, much like a "skeleton" in Human Beings. Designing an airframe from scratch involves important concepts of physics, aerodynamics, materials engineering and manufacturing techniques to achieve certain performance, reliability and cost criteria. We have designed the airframe of our Quadcopter with 258g of mass and made of GPR (Glass-Reinforced-Plastic), possessing a cage-like structure in its center that will offer extra protection to the electronics.

This particular detail may prove itself very useful when it comes to the test flight stage, when accidents are more likely to happen.

Figure 3   Top and the bottom plate.

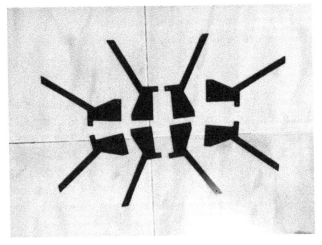

Figure 4   Quadcopter legs.

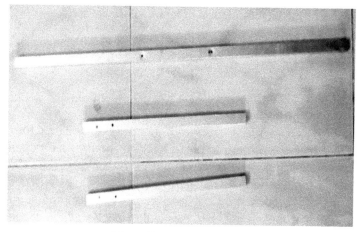

Figure 5 Aluminum frame.

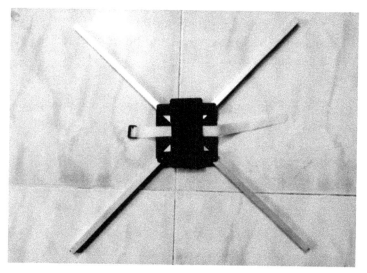

Figure 6 Quad-X frame.

2.2 Propellers

The typical behavior of a propeller can be defined by three parameters:

- Thrust coefficient c_T;
- Power coefficient c_P;
- Propeller radius r.

These parameters allow the calculation of a propeller's thrust T:

$$T = \frac{c_T 4\rho r^4 \omega^2}{\pi^2} \tag{1}$$

and power

$$P_P := \frac{c_P 4\rho r^5 \omega^3}{\pi^3} \tag{2}$$

where ω is the propeller angular speed and ρ the density of air. These formulas show that both thrust and power increase greatly with propeller's diameter. If the diameter is big enough, then it should be possible to get sufficient thrust while demanding low rotational speed of the propeller. Consequently, the motor driving the propeller will have lower power consumption, giving the Quadcopter higher flight autonomy. Available models of counter rotating propellers are scarce in the market of radio controlled aircrafts.

The "EPP1045" (see Figure 7), a propeller with a diameter of 10" (25.4 cm) and weighting 23 g, presented itself as a possible candidate for implementation in the Quadcopter. To check its compatibility with the project requisites it is necessary to calculate the respective thrust and power coefficient. We can extract the mean thrust and power co-efficient by using Equations (1) and (2):

Figure 7   EPP1045 propellers.

$$c_T = 0.1154 \tag{3}$$

$$c_P = 0.0743 \tag{4}$$

In reality, neither the thrust nor the power co-efficient are constant values, they are both functions of the advance ratio J:

$$J = u_0 n D_P \tag{5}$$

where u_0 is the aircraft flight velocity, n the propeller's speed in revolutions per second and finally D_P is the propeller diameter.

However, when observing the characteristic curves for both these coefficient as shown in Figures 8 and 9, it is clear that when an aircraft's flight velocity is very low (e.g. in a constant altitude hover) the advance ratio is almost zero and the two co-efficient can be approximated as constants, which is the current case, for at this point there is no interest in achieving high translation velocity. Assuming the Quadcopter's maximum weight is 9.81 N (1 kg) and that we have four propellers, it is mandatory that each propeller is able to provide at least 2.45 N (1/4 of the Quadcopter weight) in order to achieve lift-off. Taking this data into consideration leads us to wonder about the minimum propeller rotational speed involved, as well as the magnitude of the power required for flight. Figure 10 helps us with some of these questions. It is observed that a propeller will have to achieve approximately 412 rad.s^{-1},

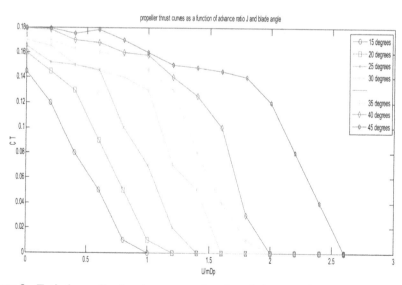

Figure 8 Typical propeller thrust curves as a function of advance ratio J and blade angle [8].

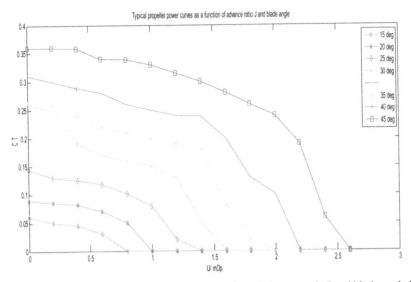

Figure 9   Typical propeller power curves as a function of advance ratio J and blade angle [8].

which is equivalent to 3934 revolutions per minute, to provide the minimum 2.45 N required for lift-off [6].

The respective propeller power is 26 W. After this short analysis we can state that the EPP1045 propellers are suitable for implementation in the Quadcopter prototype.

3 Multi-Wii Configuration

The Multi Wii Copter is an open source software which controls the RC Platform and also compatible with the hardware boards and sensors.

The first and most famous setup is the association of a Wii Motion Plus and a Arduino pro mini board. The MultiWii 328P is a gyro/accelerometer based flight controller that is loaded with features. This version of the MultiWii supports DSM2 satellite receiver functionality. With expandability options and full programmability, this device can control just about any type of aircraft. This is the ideal flight controller for your multi-rotor aircraft. The pin diagram for MultiWii 328P is given in Figure 11.

3.1 A Gimbals

A gimbals is a pivoted support that allows the rotation of an object about a single axis. A set of three gimbals, one mounted on the other with orthogonal

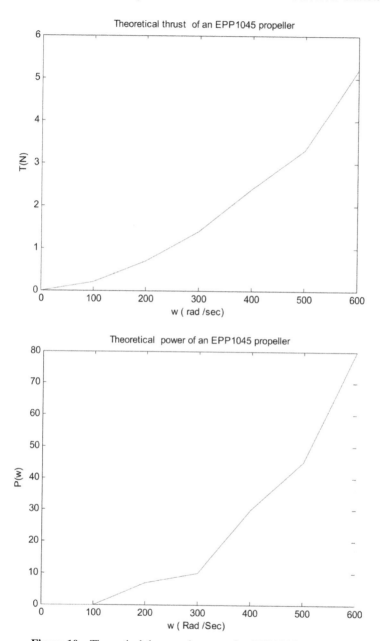

Figure 10   Theoretical thrust and power of an EPP1045 propeller.

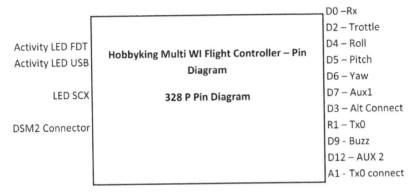

AO – CAM Pitch A1 – CAM Roll A2 –CAM Trigger   D3, D10 – Motor Connect

Figure 11   Pin diagram of MultiWii 328P.

pivot axes, may be used to allow an object mounted on the innermost gimbal to remain independent of the rotation of its support. For example, on a ship, the gyroscopes, shipboard compasses, stoves, and even drink holders typically use gimbals to keep them upright with respect to the horizon despite the ship's pitching and rolling. When associated with an accelerometer, MultiWii is able to drive 2 servos for PITCH and ROLL gimbal system adjustment. The gimbal can also be adjusted via 2 RC channels.

3.2 B ITG3205 Triple Axis Gyro [9]

This is a breakout board for InvenSense's ITG-3205, a groundbreaking triple-axis, digital output gyroscope. The ITG-3205 features three 16-bit analog-to-digital converters (ADCs) for digitizing the gyro outputs, a user-selectable internal low-pass filter bandwidth, and a Fast-Mode I2C (400kHz) interface. Additional features include an embedded temperature sensor and a 2% accurate internal oscillator. The ITG-3205 can be powered at anywhere between 2.1 and 3.6 V. For power supply flexibility, the ITG-3205 has a separate VLOGIC reference pin (labeled VIO), in addition to its analog supply pin (VDD) which sets the logic levels of its serial interface. The VLOGIC voltage may be anywhere from 1.71 V min to VDD max. For general use, VLOGIC can be tied to VCC. The normal operating current of the sensor is just 6.5 mA.

Communication with the ITG-3205 is achieved over a two-wire (I2C) interface. The sensor also features a interrupt output, and an optional clock input.

A jumper on the top of the board allows you to easily select the I2C address, by pulling the AD_0 pin to either VCC or GND; If CLKIN pin is not used jumper shoul be shorted on the bottom of the board to tie it to GND.

3.3 C BMA180 Accelerometer [9]

This is a breakout board for Bosch's BMA180 three-axis, ultra-high performance digital accelerometer. The BMA180 provides a digital 14-bit output signal via a 4-wire SPI or I2C interface. The full-scale measurement range can be set to ±1 g, 1.5 g, 2 g, 3 g, 4 g, 8 g or 16 g. Other features include programmable wake-up, low-g and high-g detection, tap sensing, slope detection, and self-test capability. The sensor also has two operating modes: low-noise and low-power.

This breadboard friendly board breaks out every pin of the BMA180 to an 8-pin, 0.1" pitch header. The board doesn't have any on-board regulation, so the provided voltage should be between 1.62 and 3.6 V for VDD and 1.2 to 3.6 V for VDDIO. The sensor will typically only consume 650 uA in standard mode.

3.4 D BMP085 Barometer [9]

This precision sensor from Bosch is the best low-cost sensing solution for measuring barometric pressure and temperature. Because pressure changes with altitude you can also use it as an altimeter! The sensor is soldered onto a PCB with a 3.3 V regulator, I2C level shifter and pull-up resistors on the I2C pins.

3.5 E HMC5883L Magnetometer [9]

The Honeywell HMC5883L is a surface-mount, multi-chip module designed for low-field magnetic sensing with a digital interface for applications such as low-cost compassing and magnetometer. The HMC5883L includes our state-of-the-art, high-resolution HMC118X series magneto-resistive sensors plus an ASIC containing amplification, automatic degaussing strap drivers, offset cancellation, and a 12-bit ADC that enables 1° to 2° compass heading accuracy. The I2C serial bus allows for easy interface. The HMC5883L is a 3.0 × 3.0 × 0.9 mm surface mount 16-pin leadless chip carrier (LCC). Applications for the HMC5883L include Mobile Phones, Netbooks, Consumer Electronics, Auto Navigation Systems, and Personal Navigation Devices.

The HMC5883L utilizes Honeywell's Anisotropic Magneto Resistive (AMR) technology that provides advantages over other magnetic sensor technologies. These anisotropic, directional sensors feature precision in-axis sensitivity and linearity.

4 Graphical User Interface

Java language is used to code in Linux platform. GUI is developed for graphical visualization of sensors, processing control of motors in Quadcopter utilizing RC Signalling.

Figure 12 shows the Multi Wii Simulation for Quadcopter with GUI. It gives flying path, speed, latitude and longitude.

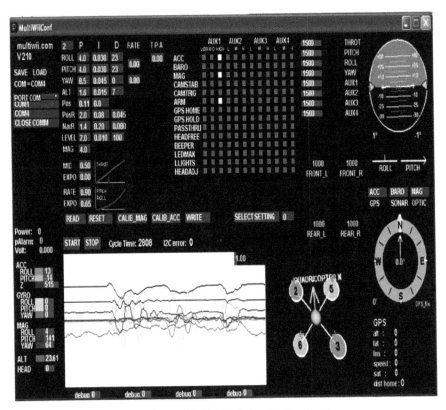

Figure 12   MultiWii Simulation for Quadcopter.

5 Android Application (SSM)

Android application is developed for measuring the signal strength through Quad Phone. Android app created in the phone has user interface with the following parameters are created.

- IMEI Number
- Cell ID
- Signal Strength in dBm
- EVDO Value
- SNR Value
- Button to fetch Details

Once this button is clicked all the details are displayed on the screen. Android app requires minimum of 1 Mb memory requirement. All these contents are visible in the User Interface of the SSM application on the quad phone, which is placed over the Quadcopter as shown in Figures 13 and 14.

Figure 13 Mobile version of the BBQ application.

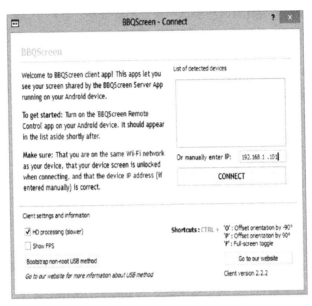

Figure 14   PC version of the BBQ client to view the phone screen on PC.

In order to view the signal strength displayed on the screen of the quad phone in PC, a new application BBQ is used. This BBQ requires the client software to be installed in the PC which is used to view the screen of the mobile phone [3]. BBQ is installed with the demo shown in Figure 13. Now the Quad Phone on the Quadcopter and the PC at the service end are connected to Wi-fi.

When BBQ is started it connects to IP automatically and this IP is connected to Mobile phone. Prior to this IP is given in BBQ software as shown in Figure 14 and thus a wireless connection between the PC and the quad phone is created. Using this wireless connection the screen of the quad phone can be viewed on the monitor of PC. Figure 15 shows the final picture of Quadcopter with Quad Phone. Figures 16(a) and 16(b) are screen shot of BBQ for measuring the signal strength.

6 Results and Conclusion

In this project, the development of the hardware and software framework necessary is undertaken to enable the Quadcopter to fly autonomously. The associated mechanical and electrical hardware is assembled and tested for

Figure 15   Final picture of the proposed Quadcopter.

its viability. The efficient design of Quadcopter housing the Quad Phone is utilized to measure the Signal Strength, SNR and and also to display the IMEI, cell ID and EVDO values associated with the Quad Phone. On analysis it is found that the proposed method of Quadcopter design developed provides a document that clearly and precisely outlines the steps necessary to assemble and fly the Quadcopter. With the implemented control scheme, the Quadcopter is able to hover autonomously and perform step movements in all directions. The experiments for several testing flying session have been performed for tuning the weight matrices of the controller and to carry out performance tests.

The main objective of this paper is to measure the radiated signal strength from a single Base Transceiver Station. This is achieved by mounting Quad Phone on a Quadcopter. An android application named "Signal Strength Monitoring (SSM)" is developed to monitor the power radiations emitted by each band i.e CDMA, GSM, HSPA, LTE in the communication protocol. So the mobile phone incorporating such an application is mounted on a Quadcopter to record the power radiation levels at several points around a single tower. Two signal strength values of 14 dbm and 11 dbm is noted for two different field tests carried out with SNR of −1 and EVDO value of −120. The Signal

(a)

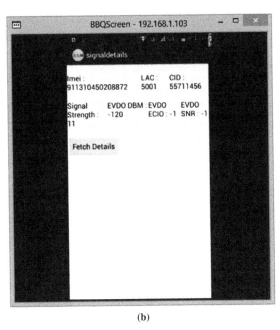

(b)

Figure 16   (a) Screenshots of the SSM Application over cell phone, (b) Screenshot of the application shared over the laptop.

Strength that is displayed by the Android Application (SSM) is successfully shared over a laptop using BBQ Software using the Wi-Fi connection and IP address.

The above design encompassing a Quad Phone within the Quadcopter could pave way for undertaking Electromagnetic Pollution Index surveys within sensitive zones, City Malls, Railway Stations, Hospital zones and Airport restricted areas etc., [5]. This technique would carve out a distinct possibility for monitoring the Electromagnetic radiation in a residential complex like multi storied buildings/flats which are directly in the line of sight of the radiating tower antennas catering to various service providers. The combination of the Quadcopter and the Quad Phone would enable auditing of the Electromagnetic Radiation (ER) and subsequently determine the Electromagnetic Pollution Index (EPI) from the delineated pockets of pollution regions.

References

[1] Mary, C. L. C. Totu, L. C., and Koldbæk, S. (2010). *Modelling and Control of Autonomous Quad-Rotor*. Aalborg: Aalborg University.

[2] Jeremia, S., Kuantanna, E., and Pangaribuan, J. (2012). "Design and construction of remote-controlled quad-copter based on STC12C-5624AD," in *Proceeding of the System Engineering and Technology (ICSET), 2012 International Conference*, 1–6.

[3] Malathi, S., TirumalaRao, G., Rajeswer Rao, G., (2013). A prediction model for electromagnetic pollution index of multi system base stations. *Int. J. Eng. Res. Technol.* 2, 12.

[4] Mattar, N. A. B., Razak, M. R. B. A., Murat, Z. B. H., Khadri, N. B., and Rani, H. N. B. H. M. (2002). Measuring and analyzing the signal strength for Celcom GSM [019] and Maxis [012] in UiTM Shah Alam campus. *Res. Dev.* 2002, 489–493.

[5] Prithiviraj, V., Cmde, J. J. N., Avadhanulu, J. (2012). "Electromagnetic pollution index- a keyattribute of green mobile communications," in *Proceeding of the Green Technologies Conference, 2012 IEEE*, Washington, DC, 1–4.

[6] Zhang, Y., Tianjin, K., Xian, B., Yin, Q., and Liu, Y. (2012). "Autonomous control system for the quadrotor unmanned aerial vehicle," in *Proceeding of the Control Conference (CCC)*, Nanjing, 4862–4867.

[7] Jorge, M., and Brito, D. (2009). "Quadrotor prototype", in *Instituto Superior Técnico*, Portugal.

[8] Prof. Z. S. *Spakovszky, Unified: Thermodynamics and Propulsion*. Available at: https://http://web.mit.edu/16.unified/www/FALL/thermodynamics/notes/node86.html

[9] Available at: https://http://www.geeetech.com/wiki

Biographies

N. P. Kumar obtained his Bachelor of Engineering in Electronics and Communication Engineering from Rajalakshmi Institute of Technology, Chennai in 2014. Currently he is working as Backup Administrator. His areas of interests are Electronics & Circuits, Digital Circuitry and Robotics.

A. R. Kumar has completed his Bachelor of Engineering (Electronics and Communication Engineering) in 2014 from Rajalakshmi Institute of Technology, Chennai. Currently he is working as Assistant System Engineer in Tata Consultancy Services. His areas of interest include Robotics, Digital Electronics and Mobile Communication.

S. Anand obtained his Bachelor of Engineering in Electronics and Communication Engineering from Rajalakshmi Institute of Technology, Chennai in

2014. Currently he is working as Engineer-Trainee. His areas of interests include Networking and Digital Circuitry.

E. N. Ganesh received M.Tech. degree in Electrical Engineering from IIT Madras, Ph.D. from JNTU Hyderbad. He has over 20 years of academic experience and now working as Dean (Research and Innovation) at Rajalakshmi Institute of Technology. His area of interests is Nanoelectronics, Robotics and Hyperspectral Image Processing.

V. Prithiviraj received M.S. degree in Electrical Engineering from IIT Madras., Ph.D. in Electronics and Electrical Communication Engineering from IIT Kharagpur. He is working as Principal Rajalakshmi Institute of Technology from May 2013. He has over 3 decades of teaching experience and 12 years of Research & Development Experience between the two IITs in the field of RF & Microwave Engineering. His areas of interest include Broadband and Wireless Communication, Telemedicine, e-Governance and Internet of Things.

Lyapunov Optimization Based Cross Layer Approach for Green Cellular Network

L. Senthilkumar[1,*], M. Meenakshi[2] and J. Vasantha Kumar[3]

[1]*Research Scholar, Anna University, Chennai, India*
[2]*Professor, Anna University, Chennai, India*
[3]*Master graduate Student, Anna University, Chennai, India*
E-mail: senthilkumar1@live.com; meena68@annauniv.edu;
j.vasanth489@gmail.com
*Corresponding Author

Received January 2015; Accepted March 2015;
Published April 2016

Abstract

Cross layer techniques are conventionally proposed for improving the network performance and recently many models have been proposed for improving energy efficiency. However most of these models have not considered all the fundamental Quality of service requirements along with the energy efficiency. Quality of Service and Queue Stability affect the energy consumption and network performance in every time slot of network operation. So an adaptive model is necessary to guarantee the Quality of service and Queue stability along with energy consumption. The proposed model applies the stochastic drift plus penalty method to optimize energy efficiency subject to Quality of Service and Queue stability constraints. The optimization technique in the proposed model does not require knowledge of channel density function. The simulation results suggest the improved energy efficiency of the proposed model.

Keywords: Cross layer, Quality of Service, stochastic drift plus penalty method, Energy efficiency, Queue Stability.

Journal of Green Engineering, Vol. 5, 129–150.
doi: 10.13052/jge1904-4720.523

1 Introduction

Cross layer design based network performance improvement has been an evolving strategy in recent times. Though many adaptation schemes are deployed in different OSI layers, the lack of coordination among them makes the overall performance of the system non-optimal. Only proper coordination across layers can benefit the system to achieve Quality of Service (QoS) with optimized goals across layers. For some applications, the packet arrival rate at the transmission buffer is continuous, while for some other applications, the packet arrival is quite bursty in nature. Therefore, if the packet scheduler at the lower layer does not utilize the traffic information of the application it is dealing with, it may cause excessive delay (and buffer overflow when the buffering capacity is limited) and/or excessive power consumption. An intelligent packet scheduler should be able to adjust the transmission rate at the physical layer depending not only on the channel gain, but also on the buffer while satisfying the QoS requirements on delay, overflow and packet error rate. For example, when packet delay is relatively less important than transmission power, the scheduler should not hurry up transmission by using a higher power level in bad channel conditions when the buffer has relatively fewer packets. It can wait for a better channel condition.

This technique achieves two goals: it satisfies the packet error rate, delay, and overflow requirement, and it does so with the lowest possible transmission power. In future green radio networks, the scheduler will have to apply similar techniques to save energy. In this paper, we show how joint optimization can be used in an intelligent scheduler to reduce energy consumption.

Among all the cross-layer adaptation techniques, the rate and power adaptation techniques at the physical layer are the most important ones for green radio network design, since they minimize transmission power based on upper layer information. Therefore, without loss of generality, in this paper, we concentrate on the power minimization issue that is of particular importance for green radio networks. We show how the transmitter power can be saved using cross-layer optimal policies, where the rate and power at the physical layer are adjusted to minimize power with particular delay, packet error rate and overflow requirements striking a balance between "green need" and service requirements.

Objective of our work is to design a cross-layer based cellular architecture that maximizes the system performance with the energy and QoS constraints. In order to allocate the resources in the system, the scheduler utilizes both channel and queue information. Therefore, in every time-slot n, the scheduler monitors the states of the traffic, buffer and channel, and then allocates resources dynamically. The concerned problem falls under the category of stochastic dynamic programming problems. To solve this dynamic problem, we use Lyaponuv drift plus penalty algorithm where in the number of packets to be transmitted in each time slot is determined for transmission over fading channels considering both the physical layer and the data-link layer optimization goals. At the physical layer, our goal is to optimize the transmission power while satisfying a particular bit error rate (BER) requirement. On the other hand, at the data-link layer, our goal is to optimize the delay and packet loss due to overflow. Overall, the cross-layer approach is shown to be effective in conserving the energy of the system while satisfying the QoS requirements.

1.1 Related Works

Energy-efficient cross-layer optimized techniques and designs were a major research attention in the last decade among wireless researchers working in different networks and protocol stacks.

In [1], the authors have presented a cross-layer based design technique that evaluate the adapive policy based on both the physical layer and the data-link layer information. In this scheme, delay, overflow rate and BER are guaranteed precisely for all traffic arrival rates. Also they have achieved significant system-level throughput gain using cross-layer adaptation policy compared with single-layer channel-dependent policy.

A study of energy efficiency of emerging rural-area networks based on flexible wireless communication is presented in [5]. Authors have given clear approaches to energy efficient PHY parameter adjustment and also added into consideration the notion of physically achievable modulation and coding schemes. Neural network based cross-layer adaptation scheme was proposed in [11], which improves QoS by online adapting media access control (MAC) layer parameters depending on the application layer QoS requirements and physical layer channel conditions.

The problem of optimal rate control scheme which uses dynamic programming based optimization technique is proposed in [12] for wireless networks with Rayleigh fading. Energy-efficient transmission techniques are discussed in [7], for Rayleigh fading networks, where the authors show how to map the wireless fading channel to the upper layers parameters for cross-layer design. In [14], an energy-efficient cross-layer design is studied for MIMO downlink. In [4], authors have related the error vector magnitude (EVM), bit error rate (BER) and signal to noise ratio (SNR). They also present the fact that with such relationship it would be possible to predict/substitute EVM in places of BER or even SNR.

Energy-efficient operation modes in wireless sensor networks are studied in [13] based on cross layer design techniques over Rayleigh fading channels using a discrete-time queuing model and a three-dimensional nonlinear integer programming technique. The authors in [10] have shown the joint optimization of the physical layer and data link layer parameters (e.g. modulation order, packet size, and retransmission limit). The problem of optimal trade-off between average power and average delay for a single user communicating over a memory-less block fading channel using information-theoretic concepts is investigated in [9].

1.2 Paper Organization

This paper is organized as follows. Section 2 describes system model including traffic and buffer models. Section 3 describes computation of transition matrix in WARP test-bed. In Section 4, we formulate the cross-layer design problem and discuss cost functions and constraints. Section 5 provides methodology of finding optimal policies and optimal costs for different objectives or QoS requirements using Lyapunov drift penalty algorithm. We discuss the results in Section 6 and conclude in Section 7.

2 System Model

Figure 1 shows the wireless transmission link with single transmitter and single receiver. This model is similar to the cellular networks where we focus only on the transmission from a base station (BS) to a single mobile station (MS). Transmission time frame is divided into discrete set of time-slots. The processing units are packet for higher layer and block for the physical layer. A downlink (or uplink) transmission block contains symbols and a packet contains the information bits. Packets from the higher layer application are stored in a finite size queue or buffer at the transmitter. Adaptive modulator (AM) is employed which chooses the modulation scheme based on the information of channel state, queue state, BER and incoming traffic state. The AM unit takes packets from the buffer and modulates it with the chosen modulation scheme into symbols for transmission over channel.

Let T_B denote the time-slot duration and hence the time-slot rate is $R_B = 1/T_B$ (time-slots/second). Let b(n) denote departure rate of the buffer, that is, number of packets taken from the buffer for transmission over the nth time

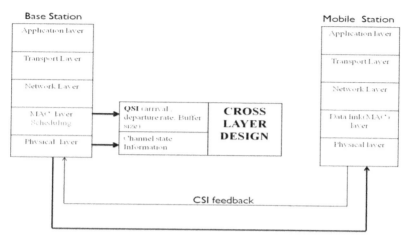

Figure 1   System model.

slot and a(n) is the arrival rate, that is, packet coming to the transmitter buffer from the upper layer application consisting of bits per slot. Assume that the block duration is equal to N number of discrete time-slots.

S denotes the possible system states. The job of a PHY-layer scheduler is to find the control action $\mu \in \{S_1, S_2, S_3, S_4, S_5, S_6\}$ for all the time-slot S_n; $n = 1, 2, \cdots, N$, where μ is the set of available actions and N (in number of slots) is the duration of communications. Our goal is to find an optimal policy μ so that it translates the state into a corresponding optimal action, $\mu(s)$, We will discuss later how the action, which corresponds to transmission rate and/or power of the problem, can be selected when adaptation is made with only the physical layer, and also with cross-layer variables.

2.1 Buffer Modeling

In our proposed model we assume that buffer which stores the higher layer packets, has a finite size of B. Since arrival rate and departure rate are random, the buffer may be empty, partially empty or may be fully occupied at different time slots. If the buffer is completely occupied, then overflow occurs, that is, some packets will be dropped. Therefore, in this paper our goal is to allocate resource by considering the issues of both the queuing delay and the packet overflow. Let Q(n) denote buffer occupancy at time slot n, therefore, the state space of the buffer's packet occupancy can be expressed as $Q(n) = \{0, 1, 2 \cdots, B\}$. Higher layer traffic produce arrival rate a(n) and departing packet rate b(n) is a function of channel and power.

2.2 Traffic Modeling

Usually, wireless network traffic is bursty, correlated and randomly varying. The Markov modulated Poisson process (MMPP) model, where in, at any state, the incoming traffic is Poisson distributed. A packets may arrive according to the Poisson distribution with average arrival rate λ_i (packets/time-slot). In each time-slot, the transmitter selects b(n) packets for transmission over the wireless channel. In each time slot modulation schemes are chosen based on the cost function of the algorithm. Since the number of arrived packets a(n) and the number of packets chosen for transmission b(n) are randomly varying, the buffer occupancy fluctuates between 0 to B, where B is the storage capacity of the buffer. The buffer state at time-slot n can be given as

$$Q(n) = Q(n-1) + a(n) - b(n) \tag{1}$$

3 Warp Testbed

In the present work, the Channel distribution is found by WARP [16] test-bed. Total received signal strength is dependent both on distance and fading. We have assumed that the distance remains unaltered during the time period of interest, and hence we can just rely on fading to capture the variations in signal strength. However in situations where the above premise does not hold true one can combine this fading-based Markov chain model with a mobility to model signal strength fluctuations.

The channel state partitioning can be done in different ways, but the equal probability method, where all the channel states have the same stationary probability, is the most popular in literature, because it offers a good trade-off between the simplicity and the accuracy for modeling a wireless fading channel. We denote the channel states by $C_k = \{C_1, C_2, C_3, C_4, C_5, C_6\}$, where the state is said to be in C_k when the gain lies between γ_{k-1} and γ_k as shown in Table 1.

Table 1   Channel states

Channel State	SNR(dB)
C_1	3.0–5.0
C_2	5.0–8.0
C_3	8.0–10.5
C_4	10.5–14.0
C_5	14.0–18.0
C_6	Greater than 18.0

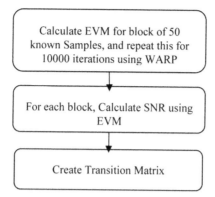

Figure 2 Flow chart for determining the transition matrix.

3.1 Transition Matrix

The transition matrix of SNR variations can be determined by collecting received signal strength measurements and calculating the EVM values using the WARP Test-bed and then determining the transitions from one state to the other.

The transition probabilities $P_{ci,cj}$; c_i, c_j $\forall$ C_k were determined based on SNR variations which are obtained directly from EVM. The state of the channel can be estimated at the receiver and the information can be fed back to the transmitter. When the perfect channel state information is available at the transmitter before the transmission decision is taken, we usually refer to the channel as fully observable.

We obtain the Markov chain transition matrix via an empirical approach, in which the Markov chain transition matrix is calculated by directly measuring the changes in signal strength. Figure 2 shows the flow chart for obtaining the transition matrix.

The transition matrix can be determined by performing signal strength measurements at the receiver for experiments conducted over fading channel. The first step in framing transition matrix is to calculate EVM values for each block with assumption that each block consist N samples. EVM can be calculated by using following expression [4],

$$
EVM = \left[\frac{\frac{1}{T} \sum_{t=1}^{T} |I_t - I_{0,n}|^2 + |Q_t - Q_{0,n}|^2}{\frac{1}{N} \sum_{n=1}^{N} \left[I_{o,n}^2 + Q_{o,n}^2 \right]} \right]^{\frac{1}{2}} \tag{2}
$$

$$SNR = \left[\frac{\frac{1}{T} \sum_{n=1}^{T} |I_t^2 + Q_t^2|^2}{\frac{1}{N} \sum_{n=1}^{T} \left[n_{I,t}^2 + n_{Q,t}^2 \right]} \right] \tag{3}$$

where

I_t, Q_t – Received symbol at t'th instant

$I_{0,t}, Q_{0,t}$ – Transmitted symbol at t'th instant

$I_{0,n}, Q_{0,n}$ – N Unique ideal Constellation Points

From above equations it is clear that, SNR is inversely proportional to square of EVM.

$$\text{SNR} \approx 1/\text{EVM}^2 \tag{4}$$

Therefore, we have the sequence of SNR values from which we can obtain SNR transition matrix using hidden Markov model.

The number of transitions from each state to the others is determined by observing the sequence of states. For example, suppose there are 6 states in all and that the sequence of states is {......2, 4, 6, 2, 4 ......}. The subsequence {2, 4} means that we increment the number of transitions from state 2 to state 4 by one. The next transitions are from states 4 to 6, 6 to 2 followed by another transition from 2 to 4. Once all the transitions have been considered, we use the relative values of the number of transitions from state i to state j for all states j, to determine the empirical transition probabilities from state i to all states j, P_{ij}.

A stochastic matrix (also termed transition matrix), is a matrix used to describe the transition of SNR between various states in Rayleigh Channel. The obtained transition matrix is

$$P_{ij} = \begin{array}{c} \\ S_1 \\ S_2 \\ S_3 \\ S_4 \\ S_5 \\ S_6 \end{array} \begin{bmatrix} S_1 & S_2 & S_3 & S_4 & S_5 & S_6 \\ 0.3651 & 0.2262 & 0.1710 & 0.1478 & 0.0817 & 0.0082 \\ 0.2286 & 0.2388 & 0.2105 & 0.1729 & 0.1289 & 0.0204 \\ 0.1390 & 0.2548 & 0.2221 & 0.1830 & 0.1617 & 0.0395 \\ 0.0592 & 0.2339 & 0.2294 & 0.2244 & 0.1893 & 0.0638 \\ 0.0092 & 0.1564 & 0.2669 & 0.2368 & 0.2215 & 0.1092 \\ 0.0001 & 0.0527 & 0.2109 & 0.3018 & 0.2491 & 0.1855 \end{bmatrix}$$

Table 2 Possible actions based on channel states

Actions	MCS
S_1	BPSK(1/2)
S_2	QPSK(1/2)
S_3	QPSK(3/4)
S_4	16QAM(1/2)
S_5	16QAM(3/4)
S_6	64QAM(2/3)

We observe from the above that the total variation is small, which implies that the distributions are close to each other. i.e., probability of SNR being in same state for next slot is higher than probability of changing to adjacent state in next time slot.

Based on the system state the corresponding cost functions are calculated and actions are taken accordingly. The possible actions are shown in Table 2.

These typical values have been used directly as possible action for traditional adaptive scheme reported in literature. In our paper, we are taking Queue State Information along with channel states to take optimum action.

4 Problem Formulation

4.1 Energy Considerations

In digital communications, as a general rule, energy consumption is lowered by either shortening transmission time or lowering transmission power. Higher bit rates lower the transmission time, but are sustainable only when the power is high enough to result in sufficient SNR. Thus, unless we allow for data to be dropped, a tradeoff between the time and the power exists. The theoretical relationship between bit rate and transmission power is given by Shannon's formula, which defines the boundary for the channel capacity. Since the formula does not provide a means to achieve the boundary bitrates, a theoretical solution can be practically infeasible. Moreover, in theory, the transmitter power is usually analyzed in isolation, while in reality the transmitter needs supporting hardware, which has non-zero power consumption.

4.2 Energy Efficiency

Assuming the popular OFDM based PHY signal, with some approximation (discarding the guard intervals), we can consider the subcarriers individually, and for each of them Shannon's formula defines the maximum achievable bitrate as:

$$R_i = W \log_2 \left(1 + \frac{\gamma_i}{\Gamma}\right) = W \log_2 \left(1 + \frac{p_{Tx,i}\, g_i}{N_o W \Gamma}\right) \tag{5}$$

where,

W represents the bandwidth occupied by a single subcarrier

γ_i represents signal-to-noise ratio

g_i channel gain

$P_{Tx,i}$ transmission power at the ith subcarrier

N_0 represents power spectral density of white Gaussian noise

Γ SNR gap

Total maximum data rate for k sub carriers is

$$R = \sum_{i=0}^{k} R_i \tag{6}$$

Total energy consumed by a bit is

$$E_{Txb} = \frac{P_{Tot}}{R} \tag{7}$$

Where, P_{Tot}–Total Power

In data transmission significant part of the energy goes to the transceiver circuit power (P_{TC}), which takes into account the consumption of device electronics, such as mixers, filters and DACs, and is bitrate independent. With a non-zero P_{TC} the energy consumption is:

$$P_{Tot} = \sum_{i=0}^{k} P_{Tx,i} + P_{TC} \tag{8}$$

The bitrate used in the calculations represents an upper bound. In physical systems the choice of MCS determines the actual bitrate. This bitrate is below the optimal for the given SNR, but is equal to the optimal for a channel with a SNR lower by a factor Γ. This factor is called the "SNR gap" and depends on the MCS used, as well as the desired bit error rate (BER). The energy per bit becomes:

$$E_{Txb} = \frac{\sum_{i=0}^{k} P_{tx,i} + P_{TC}}{\sum_{i=0}^{k} W \log_2\left(\frac{P_{Tx,i}g_i}{N_oWT}\right)} \tag{9}$$

where, g_c = coding gain

$$\Gamma = \frac{\log\left(\frac{P_e}{0.2}\right)}{g_c}$$

$$p_e = \frac{4}{\log_2 M} Q\left\{\sqrt{\frac{a_\gamma \log_2 M}{(M-1)}}\right\}$$

and P_e is Probability of error.

4.3 Problem Description and Formulation

At each time-slot, the scheduler chooses an action depending on the current system states. A decision rule denoted with μ specifies the action at time-slot n. We consider a countably infinite horizon problem, where our objective is to optimize long term average expected cost for targeted goals to be achieved. The targeted goal in this paper is to minimize the average power consumption subject to limitations on the average buffer delay and packet overflow. Let Π denotes the set of all admissible policies Π, i.e., the set of all sequences of actions $\mu = \{\mu_1, \mu_2, \cdots\}$ with $\mu:S_n \in S$. The cost function for the policy is denoted by G_p. The objective of our cross-layer adaptation problem is to find the optimal stationary policy μ so that,

$$\text{Minimize} \quad G_p = E_{tot} \tag{10}$$

$$\text{subjected to} \quad G_d \leq G_{dth}$$
$$G_e \leq G_{eth}$$
$$G_o = 0$$

where, G_{dth} and G_{eth} are the maximum allowable average delay and maximum allowable probability of error respectively. $G_P(S_n)$ is the immediate transmission power cost at time slot n for action S_n. The long-term average expected queueing delay cost, G_d and packet overflow cost, G_o can be expressed in terms of the buffer backlog, number of packets arrived per slot etc.

4.4 Cost and Constraints

4.4.1 Transmission power cost

Transmission power cost relates to the power consumed by the system in a time-slot. For targeted BER, the system has to choose the adaptive power cost to maintain the QoS. The BER requirement can be specified by the application from the higher layer. For a certain channel state and action S_i, and with a fixed specified average BER P_e for all channel states, the power cost G_p is estimated with appropriate BER expression or using with instantaneous received SNR γ.

4.4.2 Queueing delay cost

Delay is an important parameter to consider for delay sensitive networks. The maximum tolerable packet delay for a particular system depends on the QoS requirements of the application from the higher layer. Factors which causes delay is composed of buffer-queuing, encoding, propagation and decoding delay. In our proposed model, we consider only buffer delay, since the other delays are fixed. The average packet delay via Little's theorem, is expresses as follows:

$$G_o(n) = \frac{Q(n)}{a(n)} \tag{11}$$

where, a(n) is the instantaneous packet arrival in slot n and Q(n) is the number of packets present in queue at time slot n.

4.4.3 Packet overflow cost

When the mean arrival rate is higher than mean departure rate packet overflow occurs. Buffer can accommodate only, r(n) = B – Q(n) + b(n) arriving packets in the current time-slot. Now, if arriving packets a(n) in particular traffic state is larger than r(n), (a(n) – r(n)) packets will be dropped with probability 1 and we say buffer overflow has occured.

Therefore, packet overflow rate, for buffer state Q(n), traffic state f(n) and action S_n can be expressed as in [1],

$$G_o(n) = \sum_{a(n)} \phi(a(n), (B - Q(n) + b(n))) * P(a(n)) \tag{12}$$

where (x, y) is a positive difference function, which returns the difference of x and y when $x > y$, and it returns 0 when $x \le y$. P(a(n)) is the probability of arrival rate being a(n) at time slot n.

4.4.4 Delay constraint

The maximum tolerable packet delay for a particular system depends on the QoS requirements of the application at the higher layer. We should maintain the delay at each slot to be less than the maximum permissible value ($\sim$Buffer delay threshold).

$$G_D(n) \leq G_{Dth}$$
$$\text{Let } g_d = G_D(n) - G_{Dth}$$
$$\text{then } g_d \leq 0$$

and hence, we should maintain g_d as a negative value.

4.4.5 Error constraint

The BER requirement is specified by the QoS of the higher layer application. We should maintain probability of error to be less than some typical value based on QoS.

$$G_e(n) \leq G_{eth}$$
$$\text{Let } g_e = G_e(n) - G_{eth}$$
$$\text{then } g_e \leq 0$$

and hence, we should maintain g_e as a negative value.

4.4.6 Overflow constraint

We assume that the finite size buffer can hold a maximum of B packets. Since the arrival rate could be random in nature, the buffer may become full during certain time slots. If the buffer does not have enough space for all incoming packets, overflow occurs. It may require retransmission which causes increase in energy. So our goal is to maintain zero overflow.

$$g_o = G_o(n)$$
$$\text{i.e., } g_o = 0$$

5 Stochastic Optimization

5.1 Drift-Plus-Penalty Algorithm

The objective function of our problem targets at minimizing the energy consumption subject to delay, error and overflow constraints and the Drift-Plus-Penalty algorithm is used to achieve the same. Hence our problem is defined as:

Min $G_p = E_{tot}$

Subject to the following Penalty functions whose time average should be minimized.

- $g_d(n) \leq 0$
- $g(n)_e \leq 0$
- $g_o(n) = 0$

5.1.1 Virtual queue

For each constraint 'i' in $\{1, \ldots, K\}$, virtual queue with dynamics over slots n in $\{0, 1, 2, \ldots N\}$ are given as follows, [15]:

Delay:

$$Z_D[n+1] = \max\left(Z_{D(n)} + g_d(n), 0\right) \tag{13}$$

BER:

$$Z_e[n+1] = \max\left(Z_e(n) + g_e(n), 0\right) \tag{14}$$

Overflow:

$$H_o[n+1] = H_o(n) + g_D(n) \tag{15}$$

where Z_D, Z_e, H_o are Lyapunov parameters used for creating virtual queues.

By stabilizing these virtual queues ensures the time averages of the constraint functions are less than or equal to zero, and hence the desired constraints are satisfied.

5.1.2 Lyapunov function

To stabilize the queues, the Lyapunov function L(n) is defined as a measure of the total queue backlog on slot n, as:

$$L(\theta(n)) = \frac{1}{2}\sum_{k=1}^{K} Q_k(n)^2 \tag{16}$$

Squaring the queueing equation results in the following bound for each queue

$$L(\theta(n)) = \frac{1}{2}\left\{\left(Q(n)^2\right) + Z_D(n)^2 + Z_e(n)^2 + H(n)^2\right\} \tag{17}$$

5.1.3 Lyapunov drift

The Lyapunov drift given by Equation (18) is used with penalty functions to identify the control action,

$$\Delta(n) = L(n+1) - L(n) \tag{18}$$

The drift-plus-penalty algorithm takes corrective actions in every slot n to minimize the Cost function. Intuitively, taking an action that minimizes the drift alone would be beneficial in terms of queue stability but would not minimize penalty. Taking an action that minimizes the penalty alone would not necessarily stabilize the queues. Thus, taking an action to minimize the weighted sum incorporates both objectives of queue stability and penalty minimization as indicated below.

Lemma [15]

$$
\Delta\left[\theta\left(n\right)\right] + VE\left\{\frac{y_o\left(n\right)}{\theta\left(n\right)}\right\} \leq B + VE\left\{\frac{y_o\left(n\right)}{\theta\left(n\right)}\right\}
$$

$$
+ \sum_{k=1}^{K} Q_k E\left\{a_k\left(n\right) - \frac{b_k\left(n\right)}{\theta\left(n\right)}\right\} + \sum_{l=1}^{L} z_l\left(n\right) E\left\{\frac{y_l\left(n\right)}{\theta\left(n\right)}\right\}
$$

$$
+ \sum_{j=1}^{J} H_j\left(n\right) E\left\{\frac{e_j\left(n\right)}{\theta\left(n\right)}\right\} \tag{19}
$$

where

$$
B \geq + \frac{1}{2} \sum_{k=1}^{K} E\left\{a_k\left(n\right)^2 - \frac{b_k\left(n\right)^2}{\theta\left(n\right)}\right\} + \frac{1}{2} \sum_{l=1}^{L} E\left\{\frac{y_l\left(n\right)^2}{\theta\left(n\right)}\right\}
$$

$$
+ \frac{1}{2} \sum_{j=1}^{J} E\left\{\frac{e_j\left(n\right)^2}{\theta\left(n\right)}\right\} - \sum_{k=1}^{K} E\left\{b_k\left(n\right)\frac{a_k\left(n\right)}{\theta\left(n\right)}\right\} \tag{20}
$$

5.1.4 Cost function

The above lemma is used to obtain the expression for the cost function as,

$$
Cost = V * g_p(n) + (1 - V) * \{Q(n) * [a(n) - b(n)] + g_D(n) * Z_D(n)
$$

$$
+ g_e(n) * Z_e(n) + H(n) * g_o(n)\} \tag{21}
$$

where V = 0.5 states that we are giving equal importance to objective (drift function) as well as penalty function. In this work, we estimate the cost function for all six possible states and select the best out of them for each slot thereby approaching the optimized solution.

6 Simulation Results

In this work, the above described optimization approach is used considering the traffic state, buffer state and the channel state that is measured for an indoor scenario using the WARP SDR module as explained in Sections 2 and 3. The performance of the adaptation policies with respect to departure rate in relay based wireless transmission downlink system with a transmitter and a receiver is shown. This indicates how the energy of transmission (Energy per bit) varies for each time slot based on overflow and delay (QSI) and SNR (CSI).

The performance was observed for 1000 time slots to gain an understanding of the dynamics and the inter-relationships. The SNR variation and Queue backlog as function of the time slot index are shown in Figure 3 and the variation in the probability of error and transmission energy at different time slots are shown in Figure 4. From these plots we can clearly observe that whenever SNR goes low, energy consumption goes high but vice versa is not true for the same entities. This is because energy consumption not only depends on the SNR, it also depends on other constraints as defined in our problem.

It is further observed that around slot number 770, the transmitted energy is very high. This can be attributed to the increased queue backlog around that time and hence an increase in probability of error, which necessitates a corrective action.

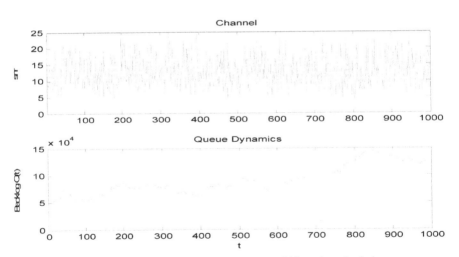

Figure 3   Lyapunov functions & Lyapunov drift vs time slot index.

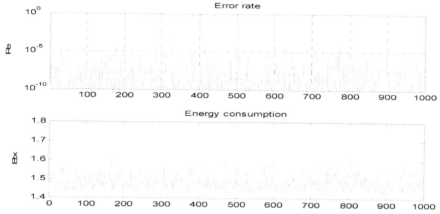

Figure 4 Variation in probability of error and transmitted energy vs time slot index.

The corresponding Lyapunov function and the Lyapunov drift at different time slot index are shown in Figure 5. We can notice that Lyapunov drift is very high at initial slots. This is because of the sudden transition of Queue backlog from lower values (zero for initial slot) to higher values.

In Figure 6, we can observe that overflow is maintained zero throughout transmission due to the corrective actions being taken at each time slot based on our optimization. It is further observed from all these performance plots that the Cost function, though dependant on many constraints, is seen to be predominantly affected by Queue backlogs, which lead buffer delays.

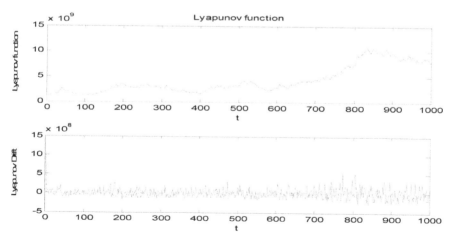

Figure 5 Lyapunov functions & lyapunov drift vs time slot index.

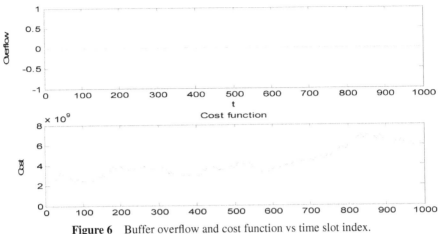

Figure 6　Buffer overflow and cost function vs time slot index.

The performance in terms of cost function and energy consumption are compared for the conventional CSI based adaptation approach and the crosslayer based approach proposed in this work and are shown in Figures 7 and 8, respectively. It can be observed that the proposed model improves the energy efficiency and also stabilize the cost function. Stability of cost function is achieved because the proposed model guarantees the stability of physical and virtual queues. The following estimates are made based on the above performances.

Total Energy Consumption (without cross layer approach) for 1000 slots = **1.4832×10^3 mW**

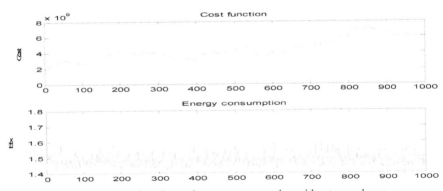

Figure 7　Cost function and energy consumption without cross layer.

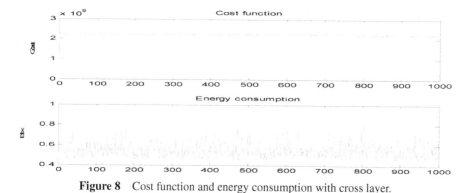

Figure 8 Cost function and energy consumption with cross layer.

Energy Consumption (with cross layer approach) for 1000 slots = **547.67 mW**

Total Cost (without cross layer approach) for 1000 slots = **4.01 $\times$ 10^{12}**

Total Cost (with cross layer approach) for 1000 slots = **2.23 $\times$ 10^{12}**

Thus the proposed approach is seen to significantly reduce the energy consumption by nearly 64% during the observation period, in comparison to the conventional adaptation strategy, in addition to stabilizing the cost function at a much reduced value.

7 Conclusion

In this paper, we have shown a possible strategy to design an intelligent packet transmission scheduler, which takes optimal transmission decisions using cross-layer information, based on Markov decision process formulations. We have discussed the method to compute the optimal policies when the channel states are perfectly observable and when they are partially observable. We have shown the benefits of a cross-layer policy over a single layer adaptation policy in terms of energy efficiency. By delaying packet transmissions in an optimal way, a huge amount of power can be saved for delay-tolerant data traffic applications. The amount of saving depends on factors such as the memory of a fading channel and the packet arrival rate. Such a cross-layer optimized packet transmission scheduling method will be a key component in future-generation green wireless networks.

References

[1] Karmokar, A. K., and Bhargava, V. K. (2009). Performance of cross-layer optimal adaptive transmission techniques over Diversity Nakagami-m fading channels. *IEEE Trans.* 57, 3640–3652.

[2] Marques, A. G., Figuera, C., Rey-Moreno, C., and Simo-Reigadas, J. (2013). Asymptotically optimal cross-layer schemes for relay networks with short-term and long-term constraints. *IEEE Trans.* 12, 333–345.

[3] Marques, A. G., Lopez-Romos, L. M., Giannakis, G. B., Ramos, J., and Caamano, A. J. (2013). Optimal cross-layer resource allocation in cellular networks using channel and queue-state information. *IEEE Trans.* 61, 2789–2807.

[4] Shafik, R. A., Shahriar Rahman, M. D., and Razibul Islam, A. H. M. (2006). "On the extended relationships among evm, ber and snr as performance metrics," in *Proceeding of the 4th International Conference on Electrical and Computer Engineering.*

[5] Pejovic, V., and Belding, E. M. (2010). "Energy-efficient communication in next generation rural-area wireless networks." in *Proceedings of the 2010 ACM Workshop on Cognitive Radio Networks (coronet' 10)*, New York, NY, 19–24.

[6] Seetharam, A., Kurose, J., Goeckel, D., and Bhanage, G. (2012). "A Markov Chain Model for Coarse Timescale Channel Variation in an 802.16e Wireless Network," in *2012 Proceedings IEEE INFOCOM.*

[7] Li, G., Fan, P., and Letaief, K. B. (2009). Rayleigh fading networks: a cross-layer way. *IEEE Trans. Commun.* 57, 520–529.

[8] Hossain, E., Bhargava, V. K., and Fettweis, G. P. (2012) *Green Radio Communication Networks.* Cambridge: Cambridge University Press.

[9] Berry, R. A., and Gallager, R. G. (2002). Communication over fading channels with delay constraints. *IEEE Trans. Inf. Theory* 48, 1135–1149.

[10] Cheng, H., and Yao, Y.-D. (2013). Link optimization for energy-constrained wireless networks with packet retransmissions. *Wiley Wireless Commun. Mobile Comput.* doi: 10.1002/wcm.996

[11] Wang, C., Lin, P. C., and Lin, T. (2006). A cross-layer adaptation scheme for improving IEEE 802.11e QoS by learning. *IEEE Trans. Neural Netw.* 17, 1661–1665.

[12] Razavilar, J., Liu, K. J. R., and Marcus, S. I. (2002). Jointly optimized bit-rate/delay control policy for wireless packet networks with fading channels. *IEEE Trans. Commun.* 50, 484–494.

[13] Lin, X.-H., Kwok, Y.-K., and Wang, H. (2012). Cross-layer design for energy efficient communication in wireless sensor networks. *Wiley Wireless Commun Mobile Comput*, 9.

[14] Dechene, D. J., and Shami, A. (2010). Energy efficient quality of service traffic scheduler for MIMO downlink SVD channels. *IEEE Trans. Wireless Commun.* 9, 3750–3761.

[15] Neely, M. J. (2010). Stochastic network optimization with application to communication and queueing system, San Rafael, CA: Morgon and Claypool Publications.

[16] WARP Project. http://warpproject.org

Biographies

L. Senthilkumar received B.E. degree in Electronics and Communication Engineering from Coimbatore Institute of Engineering and Technology, Coimbatore, India, in 2009. He successfully completed M.E. in Communication systems at College of Engineering, Anna University, Chennai, India in 2012. He is currently working toward the Ph.D. degree at College of Engineering Guindy, Anna university, Chennai, India. His current research interests include Cross-Layer design, Green optimization in telecommunication and cooperative communication.

M. Meenakshi Professor, Department of Electronics and Communication Engineering, Anna University Chennai, Guindy campus, Chennai-600025; (e-mail: meena68@annauniv.edu), India. Member of Anna University Research gate. Published 40 national and international journal papers also more than 60 national and international conference papers in the field of Optical Communication & Networks. Currently doing Research on Power optimization in small cell, Radio over fiber Networks & Communication Networks.

J. Vasantha Kumar is a M.E. Post Graduate student at the Department of Electronics and Communication Engineering, College of Engineering Guindy, Anna University, Chennai, India. He pursued his B.E. in Electronics and Communication Engineering from Veltech multi tech Dr. RR Dr. SR Engineering College, Avadi, Chennai, Tamil Nadu, India. His field of interest is wireless communication and networks and Green Communication Networks.

Environmental Hazards and Health Risks Associated with the Use of Mobile Phones

Manivannan Senthil Velmurugan

Research Scholar, D.B. Jain College, University of Madras, Chennai, India
E-mail: msvelu72@gmail.com

Received 19 January 2016; Accepted 30 March 2016;
Publication 2 May 2016

Abstract

Mobile phones are universally popular due to their convenience. Mobile phones solve problems such as interacting with the people, transfer of data through offering new channels of communication by using a device small enough to fit into one hand. On the other hand, mobile phones may be harmful to the environment and health, and waste disposal issues may be associated with its discharge of radiation. Concerns have been raised recently about the sustainability of mobile phones and its effects on people's health and the environment. The present study discusses the adverse effects associated in using mobile phones, and addresses sustainable perspectives to overcome the same.

Keywords: Mobile phone, hazards, environment, health and sustainable.

1 Introduction

Mobile phones have become an intrinsic part of most people's lives, connecting them with other people around the world. A mobile phone has several advantages, enabling communication with family, friends, and business wherever a signal is available. In addition, the 3G telephone enables users to access data; listen to music; play games; send and receive simple text messages known as short message service (SMS); access multimedia messaging services (MMS), voice, and video, as well as internet access

Journal of Green Engineering, Vol. 5, 151–174.
doi: 10.13052/jge1904-4720.524

through wireless application protocol (WAP). Even though mobile phones have several advantages, there are also significant disadvantages associated with its use. Chemical substances from mobile phones such as arsenic, lithium, cadmium, copper, lead, mercury and zinc are considered toxic. Bereketli et al, (2009), Lincoln et al. (2007) also stated that mobile phones contain a large number of hazardous substances, including antimony, arsenic, beryllium, cadmium, copper, lead, nickel, zinc and this persistent bio accumulative toxins (PBTS), have been associated with cancer and a range of reproductive, neurological, and developmental disorders. When mobiles are discarded, these toxic substances may be released or exposed from decomposing waste in landfills, contaminate the soil and seep into groundwater. Savvilotidou et al. (2014) focused on the determination of the toxic metal content of liquid crystal displays presented in various waste electrical and electronic equipment (WEEE) included mobile phones. Kang et al, (2013) explored that lithium batteries may contribute substantially to environmental pollution, and adverse human health impacts due to potentially toxic materials. Plastics are the leading chemical substance found in mobile phones followed by other miniature materials (Figure 1). Metals build-up in the soil, which can enter the food chain, and in sufficient concentrations may cause health problems. Bharodiya and Kayasth (2012), Lakshmi and Nagan (2010) explained the health hazards of manufacturing components of cell phone along with the

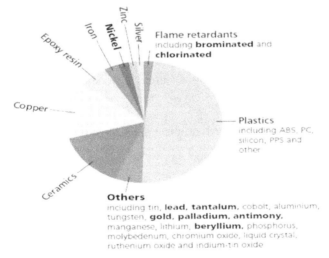

Figure 1   Chemical substances in mobile phones.

Source: Nokia, 2005.

spirit into the lives of individuals. Using mobile phones can harm the brain, and excessive use of mobile phones has been associated with dizziness. The radiations emitted from the phone are also harmful for the eardrum. Furthermore, World Health Organization (WHO, 2013) said that exposure to the radiofrequency (RF) fields emitted by mobile phones are generally 1000 times more than that emitted from base stations, and noted that research had almost exclusively conducted on the possible effects of mobile phones, such as electromagnetic interference, road traffic accidents, cancer and other health-related effects.

Moving towards a more sustainable system of mobile phone production and consumption would help to mitigate or minimize the aforementioned risks associated with mobile phones.

This paper reports the issues related to the use of mobile phones and also focuses on the aspects of sustainability in mobile phones to overcome its adverse effects. The mobile phones could have detrimental effects on people's health and the environment. The major issues related to using mobile phones are as follows:

- Lack of awareness about the health risks.
- Lack of understanding the influence on the environment.

In this paper we will first identify the associated health and environmental risks, and then examine policies and innovation strategies to reduce these risks. The analysis has a survey character, as we derive many insights from the existing literature on the problems and solutions. The literatures of the current research were fallen on mobile phones' risks in regard to energy consumption, environment and health.

1.1 Energy Consumption

The production of new mobile phones contribute to the climate change by exhausting energy and virgin materials in processes, thereby releasing greenhouse gases into the atmosphere. The United Nations Environment Programme (Kick the Habit, 2008) estimated that the manufacture of a mobile phone produces about 60 kg of CO_{2e} and using a mobile phone for a year produces about 122 kg of CO_{2e}. The evaluation of material and energy consumption of mobile phones is an important task in the end-of-life management of electronic products (Yu et al, 2010). The CO_2 emissions per subscriber in a year for a 3G system are equivalent to the emissions from driving a car for 250–380 kms or to 19–21 lts of gasoline (Nokia, 2005). Soonenschein et al. (2009) found that energy consumption for the mobile phone call averages 0–5 kWh, whereas the washing machine uses about 0–43 kWh. Besides, Groupe Speciale Mobile Association (2012) mentioned that if 10% of the world's mobile phone users turned off their chargers after use, the energy saved in one year could power 60,000 European homes. Paiano et al. (2013) examined the sustainability of the mobile phone by its energy consumption and the authors concluded that the entire mobile phone system consumes approximately 2200 GWh per year, which is equal to 0.7% of the national electricity consumption, and produces potential e-waste from end-of-life devices totaling over 11 thousand tones for the period from 2007 to 2012. Vergara et al. (2014) argued that reducing the energy consumption of wireless transmissions begins by being aware of the energy consumption characteristics of different technologies such as 3G and WiFi. Snowden (2005) stated that the energy consumption of mobile phone batteries contributes significantly to the device's environmental impact. The main impact associated with day to day mobile phone use is the power used during the charging process, and the charger represents about 7% of the life-time energy consumption of a typical mobile phone (Nokia, 2006). Most consumers are not aware of the effect of mobile phones on CO_2 production and its emissions are expected to raise 55 million metric tons due to the increase in mobile communications by 2020 (Soonenschein et al, 2009).

1.2 Environment Risks

Mobile or Cell phones are fabricated with heavy metals such as cadmium, lead, lithium, mercury and brominated flame retardants, which are used in the parts of printed circuit board (PCB), liquid crystal display (LCD), keypad, plastic casing, batteries and chargers. These substances cause severe environmental collision due to their levels of toxicity. Replacing the handsets every year, as new models become available every year, creates an unnecessary carbon footprint and hazardous waste. Robinson (2009) stated that improper disposal of waste mobile phones caused significant health effects and environmental degradation in the developing world. Mobile recycled wastes led to contamination of the soil, water, fish, and wildlife. For example, the leakage of cadmium in the battery from a single phone could contaminate 600,000 liters of water. Deloitte (2010) identified the absence of proper recycles and reuses program causing more than 7,000 tons of toxic cell phone components (i.e. more than 80 percent hazardous) to be dumped in landfills by 2012. The resulting contamination will leave far-reaching consequences to be faced by the environment and all the living forms of the earth. Tóth et al. (2012)

examined how mobile phones are used to gather environmental and health information for utilization. Unused parts of the cell phone are disposed in the environment affecting all the elements of the environment, i.e. fertility or geological structure of the land, human health, wildlife, sea and plant life (Bharodiya & Kayasth, 2012). Semiprecious metals like copper is extracted when mobile phones are recycled casually, leading to the discharge of toxicants into groundwater below and the air above (Lim & Schoenung, 2010). Lakshmi and Nagan (2010) reported that lead, a possible carcinogen can accumulate in the environment resulting in acute and chronic effects on microorganisms, plants, animals and humans, and the authors also noted when mercury makes its way into water it is transformed into methylated mercury and eventually ends up in food causing brain damage. Lithium has a high degree of chemical activity which by itself can pollute the water when exposed (Clean Up Mobile Phones, 2007). Thus, these substances may cause major crisis in the environmental by seeping from the decaying waste in landfills into ground water, contaminating the soil and eventually entering the food chain.

1.3 Health Risks

In regards to human health, the health hazards are associated with high-toxic substances released from the mobile phones. Cocosila (2007) investigated the effects of perceived health risks due to the usage of 3G mobile phones. Barnett et al. (2007) assessed the awareness of a precautionary advice contained within the Department of Health (DoH) leaflet about mobile phone health risks, and public responses to it. Lakshmi and Nagan (2010) stated that cadmium may cause lung and prostate cancer, and is toxic to the gastrointestinal tract, the kidneys, and the respiratory, cardiovascular and hormonal systems. Lead causes damage to the central and peripheral nervous systems, blood systems, and kidneys. Brominated flame retardants may increase the cancer risk of digestive and lymph systems. Thomée et al. (2011) found that increase in the frequency of mobile phone usage was associated with sleep disturbances and symptoms of depression for men and women at 1-year follow-up. Kleef et al. (2010) studied the health concerns associated with acceptance of mobile phone technology. Scientific proof is available to explain that the radiations produced by mobile phones cause severe health injury by affecting the brain of the human being (Uddin & Ferdous, 2010). Aghav (2014) notified that beyond any doubt the electromagnetic fields are harmful and its adverse effects on a human body depend upon the intensity of the cell phone frequency. Davis (2010) mentioned that the European Union sponsored REFLEX project have found significant evidence of DNA damage from signals from modern 3G phones, and also found that split samples of human sperm studied in six different national laboratories indicate poorer morphology, motility, and increased pathology in cell phone exposed samples. Acharya et al. (2013) observed that many students suffered from frequent headaches, neck pains, limb pains, back aches, and had signs of redness in their eyes and symptoms of ringing sensation in the ears or tinnitus in their ears due to continuous mobile usage on some days. Certain neurological symptoms occur due to the frequent use of mobile phones, such as depression, sadness, irritability and headaches, anxiety, loss of memory and lack of sleep. Mobile phone's electromagnetic radiations and listening to loud music will cause hearing defects. Besides, Davis (2013) has mentioned in his webpage that the exposure to radiation from cell phones may also play a vital role in the growing spate of serious problems including attention and hearing deficits, autism, behavioral changes, insomnia, tinnitus, Parkinson's disease, Alzheimer's disease and a broad array of nervous system disturbances. The conclusion arrived on the health risks associated with mobile phones reported by many researchers are given below:

Table 1   Health risks associated with mobile phones

S.No.	Authors	Conclusions
1.	Khurana et al. (2009)	Adequate epidemiologic evidence available to suggest a link between prolonged cell phone usage and the development of an ipsilateral or same sided brain tumor.
2.	Hardell et al. (2011)	An increased risk of developing a type of brain tumor (glioma) with the use of mobile or cordless phone. The risk increased with latency time and cumulative use in hours and was highest in subjects with first use before the age of 20.
3.	de Vocht et al. (2011)	Identified a small but potentially significant rise in temporal and frontal lobe brain tumors – the brain regions which are highly exposed to mobile phone radiation.
4.	Agarwal et al. (2009)	RF electromagnetic waves emitted from cell phones may lead to oxidative stress in human semen. We speculate that keeping the cell phone in a trouser pocket in talk mode may affect spermatozoa and impair male fertility.
5.	Cardis et al. (2011)	An increased risk of glioma in long-term mobile phone users with high RF exposure and of similar, but apparently much smaller risk in aquiring another type of brain tumour (meningioma).

6.	Carrubba et al. (2010)	Mobile phones trigger EP at the rate of 217 Hz during ordinary use. Chronic production of the changes in brain activity might be pertinent to the reports of health hazards among mobile phone users.
7.	Divan et al. (2012)	Cell phone use was associated with behavioral problems at 7 years of age, and this association was not limited to early users of the technology.
8.	Kesari et al. (2011)	RF electromagnetic wave from commercially available cell phones might affect the fertilizing function of spermatozoa.
9.	Kundi et al. (2011)	It cannot be dismissed from the data presented that the increase in temporal lobe malignant brain tumors (and maybe to some degree also frontal lobe tumors) is partly due to mobile phone use.
10.	Kwon et al. (2011)	Short-term mobile phone exposure can locally suppress brain energy metabolism.
11.	Levis et al. (2011)	The results from meta-analyses showed an almost doubling of the risk of brain tumours induced by long-term mobile phone use or latency.
12.	Meo et al. (2010)	Long-term exposure to mobile phone radiation leads to reduction in serum testosterone levels in Wistar albino rats. Testosterone is a primary male gender hormone, and any change in the normal levels may be devastating for reproductive and general health.
13.	Morgan et al. (2009)	Cell phones shown to cause brain tumors have potentially high public health impact.
14.	Panda et al. (2010)	Long-term and intensive mobile phone use may cause inner ear damage.
15.	Salama et al. (2010)	The pulsed RF emitted by a conventional mobile phone which was kept on a standby position, could affect the sexual behavior in the rabbit.

Aghav (2014) alarmed that most of the people were unaware about health hazard of continuous emission of radiation. Today, there is an upsurge of public concern about the possible health hazards of this new technology. The impacts of mobile phone radiation on human health have come into focus of the researchers (Anvari et al., 2013).

Apart from the health risks of mobile phone's radiation, this article also brings to light the mobile phones-related road accidents. The use of mobile phones by vehicle drivers and pedestrians will cause road accidents due to their loss of concentration. Driving while talking, texting, or using the internet distracts drivers and increases the risk of accidents. Teenagers are at the greatest risk to become victims of accidents due to cell-phone use while driving. Almost 9 in 10 teenage drivers admit to engaging in distracted driving behaviors, such as texting or talking on a cell phone. Acharya et al. (2013) emphasized that accident due to cell-phone use while driving is commonly seen on the roads today. Khan et al. (2008) confirmed that 36% of road accidents are due to the use of mobile phones while driving a vehicle. Karger (2005) also verified using a mobile phone while driving a car is significantly associated with a higher risk of vehicle collisions. Mobile phones are the commonest hand-held device with harmful effects. Therefore, it is necessary to conduct research on sustainability for defeating those issues.

2 Material and Methods

All the information and data presented in this paper were gathered from various sources of secondary data. The sources are online databases like ProQuest, Emerald, Ebsco and ScienceDirect. The online search database provided secondary data such as journals and extracts from newspapers, books and magazines. Some of the information and data were obtained from the internet search engines like Google. The research framework was developed as shown in Figure 2.

3 Result and Discussion

To reduce the environmental and health risks of mobile phones, this study identified the perspectives on sustainability such as design, manufacture, energy consumption, recycling, reusing and take back to mitigate and minimize the negative impacts of mobile phones. Although some action can be taken to mitigate these impacts, EITO (2002) also confirmed that the design, manufacture, operation and disposal of information communication technology (ICT), have an overall negative impact on the environment. The control remedies are discussed below:

3.1 Designing and Manufacturing of Mobile Phones

Mobile phones are becoming more energy-efficient and are eliminating the use of hazardous materials. Nokia (2012) suggested the integration of the concept of 'Design for the Environment' into product and technology development, and sound end-of-life practices. During product creation, they focus on energy efficiency, sustainability on the use of materials, smart packaging, and creating environmental services. This may engage people to adopt more sustainable

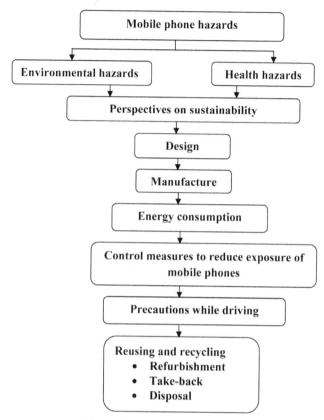

Figure 2 Research framework.

lifestyles. Environment specialist who supports the product development project should be dedicated for development of new products. They verify the implementation of legal environmental requirements and voluntary substance and materials' requirements, promote the implementation of most sustainable alternatives for material choices and energy efficiency, and provide sustainability reporting like 'Eco Profiles'. Bharodiya and Kayasth (2012) implied that the cell phone manufacturing companies have to minimize the use of hazardous metals and materials those harm the environment or may choose for those that are eco friendly. The manufacturers have to use recyclable metals and materials without emission of carbon dioxide so that there is minimum disposal on the earth. The manufacturers of cell phone have to appoint authorized distributors to sell cell phones, and the information about the authorized distributorship should be maintained by the government of respective nations. If any persons or organizations found to sell cell phones illegally without authorization from the manufacturers, strict actions should be taken. Authorized distributors should provide detailed notes about the impacts of cell phone on environment along with the user manual booklet. The government should strictly compel to follow all the policies and rules shown. If any manufacturer, distributor and customer found guilty then, appropriate fine in terms of money and/or imprisonment and, cancellation of the license may be carried out. Government can add this additional duty on the Ministry of environment to minimize or remove the impacts of cell phone on the environment. Groupe Speciale Mobile Association (2012) suggested ideal solution, from an environmental perspective, is to design a mobile phone with reuse, recycling and minimal disposal in mind. This starts with reducing energy input to the manufacture of components, substituting fewer hazardous substances (for example, the use of lead-free solder), and minimizing mixing of materials, such as metals embedded in plastics, which could be difficult to separate during recycling. Designing a phone for easy dismantling is also an important factor, as this would reduce the cost of refurbishment and recycling.

However, while the list of handset functions grows, the actual product size decreases. This produces environmental benefits through reduction in the use natural resources during manufacture, and the substitution of one device for many. Motorola is currently evaluating the use of biodegradable plastics in mobile phone covers with a view to using composting in their disposal. In order to further reduce environmental impacts, NTTDoCoMo has announced the use of recycled plastics in its new phone accessories, and Fujitsu is developing a plastic derived from corn starch. The mobile industry's pioneering switch to batteries with a higher-energy density has also produced environmental benefits. For example, power cells currently in use require fewer resources during manufacture, and avoid the use of toxic metals such as lead and cadmium. This makes the batteries much safer during the recycling and disposal processes. Mobile devices should be equipped with multiple functionalities, for instance; many include a digital camera, music player, navigation, web browser and several other features (all in one product) to help consumers reduce their own environmental footprint and avoid buying, using and charging several separate devices, when one device can be used for many different purposes.

An Integrated Product Policy project has been established to identify the improvement options that Nokia and the stakeholders could take to enhance the environmental performance of the phones. Nokia introduced the Green

Channel store, which has information about applications and services related to the environment, to raise public awareness about sustainable lifestyles, health and well-being, and social responsibility. Vodafone (2013) aimed to reduce the environmental impacts of mobile phone products and services through measures designed to empower customers to make more sustainable choices. Measures such as Eco-rating scheme launched by Vodafone, assesses the handsets on a scale between 1 and 5, with 5 being the most ethical and environmentally responsible. Manufacturers are asked to respond to more than 200 questions covering the environmental and social impacts of each phone across its lifecycle, from the mining of raw materials used to make components to consumer use and disposal, and the level of commitment of the manufacturer to managing its own impacts. This assessment is carried out annually. Vodafone concentrates on working with operators, manufacturers, suppliers, the GSMA (Groupe Spéciale Mobile Association) and the ITU (International Telecommunication Union) to develop an industry standard for measuring the environmental and social impact of mobile phones and other devices. Vodafone works with both their suppliers and customers to reduce the impacts of their products across their lifecycle. The suppliers aim to improve their sustainability of the performance by sourcing raw materials and manufacturing products thereby empowering customers to make more sustainable choices and also helping customers to achieve their sustainability goals through low carbon solutions.

3.2 Energy Consumption

In order to satisfy consumer expectations about talk and standby time, there have been significant improvements in the energy efficiency of mobile phones. Over the last 20 years, the standby operating time of a mobile phone on a single battery charge has increased from around 4 hours to up to 12 days or more, while the size of batteries has been greatly reduced. Industries have also been focusing on reducing the phone's power consumption during the charging period. However, consumers can also make an important environ-mental difference by simply switching off the phone and charger whenever possible. This has been accomplished through changes in battery chemistry and reductions in the overall energy requirements of the circuitry in mobile phone (Groupe Speciale Mobile Association, 2012). Vodafone is a signatory to the GSMA's industry-wide commitment to introduce a universal charger. This initiative aims to reduce electronic waste by eliminating the need for consumers to replace their charger when they buy a new phone. Vodafone also offered solar-powered charging solutions that can reduce environmental impacts from charging phones and extend access to reliable, renewable-energy supplies in remote areas of emerging markets. Vodafone helps consumers to make an informed choice about which mobile phone they would like to buy in order to reduce impacts from charging their phone and to recycle it when they no longer use it. For instance, Nokia estimates that if just 10% of the world's mobile phone users turned off their chargers after use, the energy saved in one year could be power 60,000 European homes. Zadok and Riikka (2010) have provided the Green Switch methodology that can be adapted to achieving positive environmental impacts; for instance, battery life increase through usage of Green Mode to reduce energy consumption during recharge of the mobile device. Nokia (2005) proposed these options for improvements in mobile phones: optimization of life-span; reduction in energy consumption and environment friendly chemicals used during component manufacture; influencing the buying, usage and disposal patterns of consumers; end-of-life, management of disposed mobile phones; reduction of energy consumption of network infrastructure; development of suitable environmental assessment methods; and development of a conducive policy environment. Hu and Kaabouch (2012) indicated that Wi-Fi consumes much less energy compared to a 3G network and using Wi-Fi as much as possible is the key to achieving more energy-efficient smartphones. Fishbein (2002) implied that mobile devices should be based on other resources, which are zinc-air technology, solar power, and muscle power in order to back up recharges rather than substitutes for batteries and adapters, and users don't need to wait for the phone to recharge.

3.3 Control Measures to Reduce Exposure of Mobile Phones

Better Health Channel (2013) suggested the following measures to reduce exposure of mobile phones:

- Choosing a mobile phone model that has a low specific absorption rate (SAR), that refers to the amount of radio frequency (RF) radiations absorbed by body tissues.
- Using a landline phone if one is available.
- Keeping your mobile phone calls short.
- Using a hands-free kit.
- Not carrying your mobile phone close to your body when it is switched on.

- Being wary of claims that protective devices or 'shields' can reduce your exposure to radio frequency (RF) radiation.

Davis (2011) also provided the following hidden safety warnings in using mobile devices:

- Don't hold a cell phone directly up to your head. Use a headset or speakerphone to talk on the phone.
- Pregnant women should keep cell phones away from their abdomen and men who wish to become fathers should never keep phones on in their pocket.
- Don't allow children to play with or use your cellphone. Older children should use a headset when talking on a cell phone.
- Turn off your wireless router at night to minimize exposure to radiation.
- Eat green vegetables and get a good night's sleep in a dark room to enhance natural repair of DNA that may have been damaged by radiation.

3.4 Precautions while Driving

For avoiding road accidents while attending to mobile phone calls, the study identified some well-being precautions for avoiding vehicle collisions. They are as follows:

- During any emergency call, stop the vehicle and then attend the call.
- Slowed the vehicle down.
- Choose a time when there was little traffic.
- Choose a time when the traffic was still or moved slowly.

Chauhan (2002) recommended the following safety measures for attending phone calls while driving a car:

- Shorter conversation will help to reduce the duration of exposure.
- Calls can be planned in such a way that longer conversations be made using ordinary land line phones.
- Minimize the conversation inside the car because the reflection from the car cavity may amplify the radiation. If this cannot be avoided the use of roof antenna would help. Use of plug-in earpiece will separate antenna further away from body/head.

3.5 Reusing and Recycling

According to the Basel Convention and the Mobile Phone Partnership Initiative (MPPI) guidance document published in November 2006, "Re-use, directly or via repair or refurbishment is usually the preferable option over recycling and disposal from an environmental perspective. Re- use can extend product life and means less environmentally damaging extraction, less energy consumption and less waste. Re-use of second-hand equipment can also often mean a lower price for products, thus increasing accessibility for more people who might not otherwise be able to afford the product". The energy and raw materials used to produce millions of new mobile phones contributes to CO_2 emissions and global warming. Mobile phones can be separated into their different components and recycled. For example, the copper, gold, lead, cadmium, silver and nickel; the gold and silver recovered can be made into jewelry. Often the batteries are first separated from the mobile phone and sorted into their various types before reprocessing by specialist recyclers. Nickel cadmium, nickel-metal hydride and lithium ion/polymer batteries have their metals recovered and reused in products such as power tools, saucepans and new batteries. The metals extracted during this process including gold, platinum, palladium and silver are put back into productive use. Phones deemed to be beyond repair or simply too old still have a residual value, and their parts may be reused. Therefore, practical and environmentally responsible methods for the recycling of end-of-life phones have to be developed in conjunction with those for electronic equipment. Groupe Speciale Mobile Association (2012) suggested that phones may be further dismantled and some parts shredded, or processed intact for material and energy recovery. Chargers, accessories and even packaging should be recycled. It is generally not economic to reuse the plastic components due to mixed grades, and the presence of dyes and other contaminants, so energy is recovered through the incineration process. In one case, energy produced from the incineration of waste materials is used to heat a village local to the recycling plant. In another, plastics are shredded and used locally in the manufacture of fence posts and pallets. The handsets, batteries, plastics and accessories may be separated according to their chemical and material composition; plastics can be recycled to make items such as traffic cones and metals are used again as good-quality raw materials. If a mobile phone is returned by a consumer and sent to a company where it may be refurbished, then it may be sold for reuse. This may not be the case if the mobile phones are not properly recycled when they reach the it end-of-life. Mobile phones have a greater chance of being reused if they are donated quickly rather than being stored. Importantly, when questioned, very few people reported that they would just throw the old phone away.

One way of reducing the negative environmental impact of smart phones is to have them recycled at the end of their lives cycle (Silveira and Chang, 2010). Vodafone (2013) encouraged customers to return their unwanted handsets and accessories to them for reuse (whenever possible) and recycling. Vodafones are also to raising awareness about their recycling programs through posters, leaflets, in-store collection points and prepaid envelopes with new handsets. Vodafone offers a buyback proactive procedure, which aims to increase the number of handsets collected for recycling, but this also has a significant commercial benefit for Vodafone. Vodafone also offers incentives for customers to keep their handsets for longer, for instance, by offering SIM-only price plans with a lower monthly subscription rate for customers who continue using their existing phone rather than upgrading. Nokia (2012) proposed that recycling programs targeted the removal of valuable materials that can be used for new products. Nokia build recycling programs by identifying safe and reliable recyclers, developing the infrastructure for reverse logistics, offering a variety of their own take-back options, and partnering with others to increase capacity to take back old mobile devices that might otherwise be headed for landfill. The research also found that used mobile phones were particularly important to people who were new to mobile telephony, those on low incomes, those under the age of 18, manual workers and non-workers. These users of used mobile phones were generally as happy with their device as users of new phones.

3.5.1 Refurbishment

Mobile phones are wiped of data, physically repaired, repackaged with new instructions and sent to suppliers. According to Groupe Speciale Mobile Association (2012), some refurbishing companies have sought environmental accreditation for proper management of the entire process. For a mobile network operator choosing a recycling partner, assurances that a partner operates good practices and, with transparency in the way they work are important. For responsible partnering companies, environmental management systems are often used to provide assurance about the proper treatment of collected phones. Collected phones must first be evaluated to determine those that are most likely to be suitable for reuse. These phones will then be subjected to a series of tests to determine suitability for reuse with or without further repair. The testing equipment and procedures are similar to those in manufacturer repair centers. Faulty parts will be replaced; batteries evaluated or exchanged, and the phone's appearance reconditioned. Particular care is needed to ensure that replacement batteries have proper internal safety circuits. All original customer information is securely erased, and the refurbished phone must meet all regulatory requirements. These steps are labor-intensive and in some cases the work is done in lower cost economies using internationally accepted health and environmental controls. Finally, the refurbished phone will be packaged for resale along with a battery, charger, and instructions. Any residual materials arising during the refurbishment process are disposed of in an environmentally sound manner. The price of a refurbished phone will vary significantly, depending upon the model type, its age and appearance. The extent of any guarantee offered by the refurbisher is another key factor. Indicative prices for one scheme are in the range of US$30–40. Mobile manufacturers incorporate measures such as material identification and easy disassembly, to make recycling easier. Many people are looking for simple phones, so when upgrading, the user may be able to find a potential user for their old phone. The user can also be a smart consumer and think twice before upgrading their mobile phone; old phone still does the job, and the cost of a new one can be saved. Nokia has actively participated in the recast of the Waste Electrical and Electronic Equipment (WEEE), that has been made to establish and develop the existing national collection networks in every country, and these networks collect and treat all electronic waste from households. This represents a big step forward to making e-waste recycling the rule, not the exception. In addition, Nokia has participated in the development of legislation concerning e-waste in countries around the world, including India, China, Kenya, Mexico and Thailand during 2011 just to name a few. Nokia takes part in collective recycling schemes with other equipment manufacturers in Europe, Canada and Australia. Nokia also engages in programs raising local recycling awareness with retailers, operators, other manufacturers and authorities around the world. Nokia's take-back and recycling programs continue to expand into new markets, ensuring mobile devices to end up in an environment safe for recycling processes.

3.5.2 Take-back

Groupe Speciale Mobile Association (2012) noted that since the 1990s, the mobile communications industry had been working in cross-sector partnerships to deliver sustainable initiatives, including used mobile phone take-back schemes that often predate national and international legislation. The industry supports handset, battery and accessory take-back in more than 40 countries. In addition, several mobile phone manufacturers have processes in place to deal with phones returned through the repair or retail outlets. In most cases, take-back schemes were established as voluntary initiatives, with self-sustaining financial structures. With some, a proportion of the revenues earned by take-back schemes are reinvested in environmental and charitable initiatives, depending on the customer culture in the individual countries. The typical price range for unsorted, used phones is in the range US$1–10. Collection and recycling programs are operated and financed by the equipment producers, while municipal collections, specified waste management sites and shops selling equipment are the main collection sites.

Experience from network operators indicate that one of the most important steps in establishing a successful take-back scheme is the incentive provided to customers. These vary depending on customers and cultural preferences but generally involve donations to charity, extra call minutes for the customer or a discount on a different phone. Nokia works to make sure consumers are aware of the channels open to them for take-back and recycling and support all safe and effective methods of mobile phone recycling. Success on take-back and recycling can be measured in three ways: the number of countries covered, the number of people reached with the recycling message in dedicated campaigns, and the weight of mobile devices and accessories recycled.

3.5.3 Disposal

If a mobile phone cannot be refurbished or if the components cannot be reused or recycled, the remaining materials are sent to an environment sound for disposal. Groupe Speciale Mobile Association (2012) reported that the remaining materials can be made insoluble at high-temperature processing so that they will not leach toxic substances into the environment, and may be safely used as a construction aggregate. In an efficient take-back program, only a tiny proportion of the materials that make up mobile phones should go for disposal (less than 10%). Consumers should be encouraged not to throw away their mobile phones with household rubbish, as the phones may end up in a landfill site. Instead, they should be deposited in a take-back scheme for refurbishment or recycling. Murphy (2008) suggested that proper disposal of old cell phones by users should involve taking it to a place that recycles cell phones or to one that will ship it off for reuse. Some companies have a take-back program and will accept the old phone when a new one is purchased from them, and also by using the old phone for a few more years you can keep it out of the cycle of waste. Bharodiya and Kayasth (2012) provided that every manufacturer has to recycle minimum 70% of metals, those used to produce the new cell phone so that only 30% of materials disposed into the environment.

Government should reserve one place to dispose, which is farthest from city to dispose the unrecyclable and hazardous parts of cell phone.

4 Conclusions

From the above discussions, sustainable strategies are needed in guiding and developing proactive customer intentions to use mobile phones with minimum risks to health and the environment. While developing awareness among customers in using mobile phones so as to minimize risks, manufacturers should develop safe measures with greenery quotes that would influence customer buying behavior and customer retention of mobiles. Companies should find and refine best sustainable solutions in improving service quality and developing the trust of users in regards to the risks caused by the use of mobile phones. The study suggested that both governments and the mobile industry should accomplish to upgrade effective regulations and legislation aspects for design, manufacture, energy consumption, recycling and reuse of mobile phones so as to mitigate and reduce the different harmful impacts.

There are a few limitations to this study, firstly, knowledge of mobile phone hazards is still accumulating, and the literature on this subject is quite limited. Secondly, this research has been conducted with secondary data, with the main aim to obtain a clear conceptual picture of the relevant problems and solution directions. However, it would be good to conduct future research using primary data on producer and consumer behaviors, perspectives and strategies to get more definite insight into mobile phone hazards and solutions to these.

References

[1] Acharya, J. P., Acharya, I., and Waghrey, D. (2013). A study on some of the common health effects of cell-phones amongst college students. *J. Commun. Health Edu.* 3, 1–4. doi: NODOI PMID:NOPMID

[2] Agarwal, A., Desai, N. R., Makker, K., Varghese, A., Mouradi, R., Sabanegh, E., et al. (2009). Effects of radiofrequency electromagnetic waves (RF-EMW) from cellular phones on human ejaculated semen: an in vitro pilot study. *Fertil. Steril.* 92, 1318–1325. doi: 10.1016/j.fertnstert.2008.08.022 PMID:18804757

[3] Aghav, S. D. (2014). Study of radiation exposure due to mobile towers and mobile phones. *Ind. Streams Res. J.* 3, 1–6. doi: NODOI PMID:NOPMID

[4] Anvari, M. M., Oliya, A., Mahdeloe, S., Harbi, T. F., Masih, M., and Bagheri, H. (2013). Environmental impacts of electromagnetic waves of mobile phones on human health. *Ann. Biol. Res.* 4, 80–84. doi: NODOI PMID:NOPMID

[5] Barnett, J., Timotijevic, L., Shepherd, R., and Senior, V. (2007). Public responses to precautionary information from the Department of Health (UK) about possible health risks from mobile phones. *Health Policy* 82, 240–250. doi: 10.1016/j.healthpol.2006.10.002 PMID:17113180

[6] Bereketli, İ., Genevois, M. E., and Ulukan, H. Z. (2009). Green Product Design for Mobile Phones. *Int. Sch. Sci. Res. Innov.* 3, 211–215. doi: NODOI PMID:17566446

[7] Better Health Channel. (2013). *Mobile Phones and your Health.* Available at: http://www.betterhealth.vic.gov.au/bhcv2/bhcarticles.nsf/pages/ Mobile_phones_and_your_health [accessed January 10, 2014].

[8] Bharodiya, A. K., and Kayasth, M. M. (2012). Impact of cell phones' life cycle on human and environment: challenges and recommendations. *J. Environ. Res. Dev.* 7, 530–536. doi: NODOI PMID:NOPMID

[9] Cardis, E., Armstrong, B. K., Bowman, J. D., Giles, G. G., Hours, M., and Krewski, D. (2011). Risk of brain tumours In relation to estimated RF dose from mobile phones—results from five Interphone Countries. *Occup. Environ. Med.* 68, 631–640. doi: 10.1136/oemed-2011-100155 PMID:21659469

[10] Carrubba, S., Frilot, C. II, Chesson, A. L. Jr., and Marino, A. A. (2010). Mobile-phone pulse triggers evoked potentials. *Neurosci. Lett.* 469, 164–168. doi: 10.1016/j.neulet.2009.11.068 PMID:19961898

[11] Chauhan, S. (2002). *Environmental and Health Hazards of Mobile Devices and Wireless Communication. CSE 6392 – Mobile Computer Systems Paper Series.* Available at: http://crystal.uta.edu/~kumar/cse 6392/termpapers/Savita_paper.pdf [accessed June 01, 2014].

[12] Clean Up Mobile Phones – Additional Information Sheet. (2007). *Mobile Phones and the Environment.* [accessed May 21, 2014] Available at: http://www.cleanup.org.au/PDF/au/additional-info-sheet_mobilephones-the-environment.pdf

[13] Cocosila, M., Turel, O., Archer, N., and Yuan, Y. (2007). Perceived health risks of 3g cell phones: do users care? *Commun. ACM* 50, 89–92. doi: 10.1145/1247001.1247026 PMID:NOPMID

[14] Davis, D. (2010). *Cell Phones: a new environmental hazard that can be reduced. Physicians for Social Responsibility, Receipt of the 1985*

Nobel Prize for Peace. Available at: http://www.psr.org/environment-and-health/environmental-health-policy-institute/responses/cell-phones-a-new-environmental-hazard-that-can-be-reduced.html [accessed June 01, 2014].

[15] Davis, D. (2013). *Cell Phones, Radiation and Your Child's Health.* Available at: http://healthychild.org/cell-phones-radiation-your-childs-health/ [accessed June 01, 2014].

[16] de Vocht, F., Burstyn, I., and Cherrie, J. W. (2011). Time trends (1998–2007) in brain cancer incidencerates in relation to mobile phone use in England. *Bioelectromagnetics* 32, 334–339. doi: 10.1002/bem.20678 PMID:NOPMID

[17] Deloitte. (2010). *Cell Phones – Need to Better Manage Growing Cell Phone Waste.* Available at: https://www.deloitte.com/assets/Dcom-CostaRica/Local%20Assets/Documents/Servicios/RCS/100830-cr_RCS _Green_phone.pdf [accessed May 20, 2014].

[18] Divan, H. A., Kheifets, L., Obel, C., and Olsen, J. (2012). Cell phone use and behavioural problems in young children. *J. Epidemiol Commun. Health* 66, 524–529. doi: 10.1136/jech.2010.115402 PMID:21138897

[19] Environment and Human Health, Inc. (2012). *The Cell Phone Problem: Technology, Exposures and Health effects.* Available at: http://www.ehhi.org/reports/cellphones/cell_phonereport_EHHI_Feb2012.pdf [accessed June 03, 2014].

[20] Fishbein, B. K. (2002). *Waste in the Wireless World: The Challenge of Cell Phones. Inform, Strategies for Better Environment.* Available at: http://informinc.org/reportpdfs/wp/WaMsteintheWirelessWorld.pdf [accessed May 25, 2014].

[21] Gsm Association. (2012). *Environmental Impact of Mobile Communications Devices.* Available at: http://www.gsma.com/publicpolicy/wp-content/uploads/2012/03/environmobiledevices.pdf [accessed May 21, 2014].

[22] Hardell, L., Carlberg, M., and Hansson, M. K. (2011). Pooled analysis of case–control studies on malignant brain tumours and the use of mobile and cordless phones including living and deceased subjects. *Int. J. Oncol.* 38, 1465–1474. doi: 10.3892/ijo.2011.947 PMID:21331446

[23] Hu, W. C., and Kaabouch, N. (2012). "Sustainable ICTs and management systems for green computing," in *Mitigating the Environmental Impact of Smartphones with Device Reuse*, Chap. 11, eds X. Li, P. Ortiz, B. Kuczenski, D. Franklin, F. T. Chong. Available at: http://www.cs.ucsb.edu/~xun/papers/reuse-ictbook12.pdf [accessed May 25, 2014].

[24] Kang, D. H. P., Chen, M., and Ogunseitan, O. A. (2013). Potential environmental and human health impacts of rechargeable lithium batteries in electronic waste. *Environ. Sci. Technol.* 47, 5495–5503. doi: 10.1021/es400614y PMID:23638841

[25] Karger, C. P. (2005). Mobile phones and health: a literature overview. *J. Med. Phys.* 15, 3–85. doi: NODOI PMID:NOPMID

[26] Kesari, K. K., Kumar, S., and Behari, J. (2011). Effects of radiofrequency electromagnetic wave exposure from cellular phones on the reproductive pattern in male Wistar rats. *Appl. Biochem. Biotechnol.* 164, 546–559. doi: 10.1007/s12010-010-9156-0 PMID:21240569

[27] Khan, A. R., Zaman, N., and Muzafar, S. (2008). Health hazards linked to using mobile cellular phones. *J. Inform. Commun. Technol.* 2, 101–108. doi: NODOI PMID:NOPMID

[28] Khurana, V. G., Teo, C., Kundi, M., Hardell, L., and Carlberg, M. (2009). Cell phones and brain tumors: a review including the long-term epidemiologic data. *Surgical Neurol.* 72, 205–214. doi: 10.1016/j.surneu.2009.01.019 PMID:19328536

[29] Kick the Habit. (2008). *A UN Guide to Climate Neutrality. United Nations Environment Programme*. Available at: http://www.grida.no/files/publi cations/kick-the-habit/kick_full_lr.pdf [accessed June 09, 2014].

[30] Kleef, E. V., Fischer, A. R. H., Khan, M., and Frewer, L. J. (2010). Risk and benefit perceptions of mobile phone and base station technology in Bangladesh. *Risk Anal.* 30, 1002–1015. doi: 10.1111/j.1539-6924.2010.01386.x PMID:20409037

[31] Kundi, M. (2011). "Comments on de Vocht et al. Time trends (1998–2007) in brain cancer incidence rates in relation to mobile phone use in England. *Bioelectromagnetics* 32, 673–674. doi: 10.1002/bem.20679 PMID:NOPMID

[32] Kwon, M. S., Vorobyev, V., Kännälä, S., Laine, M., Rinne, J. O., Toivonen, T., et al. (2011). GSM mobile phone radiation suppresses brain glucose etabolism. *J. Cerebral Blood Flow Metabol.* 31, 2293–2301. doi: 10.1038/jcbfm.2011.128 PMID:21915135

[33] Lakshmi, R., and Nagan, S. (2010). Studies on concrete containing E plastic waste. *Int. J. Environ. Sci.* 1, 282–295. doi: NODOI PMID:NOPMID

[34] Levis, A. G., Minicuci, N., Ricci, P., Gennaro, V., and Garbisa, S. (2011). Mobile phones and head tumours. The discrepancies in cause-effect relationships in the epidemiological studies – how do they arise? *Environ. Health* 10, 1–15. doi: 10.1186/1476-069X-10-59 PMID:21679472

[35] Lim, S. E., and Schoenung, J. M. (2010). Toxicity potentials from waste cellular phones, and a waste management policy integrating consumer, corporate, and government responsibilities. *Waste Manag.* 30, 1653–1660. doi: 10.1016/j.wasman.2010.04.005 PMID:20418088

[36] Lincoln, J. D., Ogunseitan, O. A., Shapiro, A. A., and Saphores, J.-D. M. (2007). Leaching assessments of hazardous materials in cellular telephones. *Environ. Sci. Technol.* 41, 2572–2578. doi: 10.1021/es0610479 PMID:17438818

[37] Meo, S. A., Al-Drees, A. M., Husain, S., Khan, M. M., and Imran, M. B. (2010). Effects of mobile phone radiation on serum testosterone in Wistar albino rats. *Saudi Med. J.* 31, 869–873. doi: NODOI PMID:20714683

[38] Morgan, L. L. (2009). Estimating the risk of brain tumors from cellphone use: Published case–control Studies. *Pathophysiology* 2, 137–147. doi: 10.1016/j.pathophys.2009.01.009 PMID:19356911

[39] Murphy, M. (2008). *Are Cell Phones Hurting the Environment?* Available at: http://voices.yahoo.com/are-cell-phones-hurting-environment-794299.html [accessed January 13, 2014].

[40] Nokia. (2005). *Integrated Product Policy Pilot Project Stage I Final Report: Life Cycle Environmental Issues of Mobile Phones.* Available at: http://ec.europa.eu/environment/ipp/pdf/nokia_mobile_05_04.pdf [accessed January 03, 2014].

[41] Nokia. (2006). *Integrated Product Policy Pilot on Mobile Phones Stage III Final Report: Evaluation of Options to Improve the Life-Cycle Environmental Performance of Mobile Phones.* Available at: http://ec.europa.eu/environment/ipp/pdf/report_02_08_06.pdf [accessed January 03, 2014].

[42] Nokia. (2012). *Nokia Sustainability Report 2011.* Available at: http://i.nokia.com/blob/view/-/1449730/data/2/-/nokia-sustainability-report-2011-pdf.pdf [accessed January 13, 2014].

[43] Paiano, A., Lagioia, G., and Cataldo, A. (2013). A critical analysis of the sustainability of mobile phone use. *Res. Conser. Recycling* 73, 162–171. doi: 10.1016/j.resconrec.2013.02.008 PMID:NOPMID

[44] Panda, N. K., Jain, R., Bakshi, J., and Munjal, S. (2010). Audiologic disturbances in long-term mobile phone users. *J. Otolaryngol. Head Neck Surg.* 39, 5–11. doi: NODOI PMID:20122338

[45] Robinson, B. H. (2009). E-waste: An assessment of global production and environmental impacts. *Sci. Total Environ.* 408, 183–191. doi: 10.1016/j.scitotenv.2009.09.044 PMID:19846207

[46] Salama, N., Kishimoto, T., Kanayama, H. O., and Kagawa, S. (2010). Effects of exposure to a mobile phone on sexual behavior in adult male rabbit: an observational study. *Int. J. Impotence Res.* 22, 127–133. doi: 10.1038/ijir.2009.57 PMID:19940851

[47] Savvilotidou, V., Hahladakis, J. N., and Gidarakos, E. (2014). Determination of toxic metals in discarded Liquid Crystal Displays (LCDs). *Res. Conserv. Recycling* 92, 108–115. doi: 10.1016/j.resconrec.2014.09.002 PMID:NOPMID

[48] Silveira, G. T. R., and Chang, S.-Y. (2010). Cell phone recycling experiences in the United States and potential recycling options in Brazil. *Waste Manag.* 30, 2278–2291. doi: 10.1016/j.wasman.2010.05.011 PMID:20554440

[49] Snowden, K. (2005). "Product design in Electronic and Electrical Engineering – mobile phones," in *Engineering for Sustainable Development Guiding Principles*, 1st Edn, eds R. Dodds and R. Venables (Westminster: The Royal Academy of Engineering).

[50] Soonenschein, M., Grabowski, S., Stenger, J., and Haas, M. (2009). *Why Go Green? A. T. KEARNEY, Inc.Marketing and Communications, Chicago, IL, U.S.A.* Available at: https://www.atkearney.com/docu ments/10192/d3e4ca69-216c-4a0e-b028-29e3fe0a8c0b [accessed May 29, 2014].

[51] Thomée, S., Härenstam, A., and Hagberg, M. (2011). Mobile phone use and stress, sleep disturbances and symptoms of depression among young adults – a prospective cohort study. *BMC Public Health* 11, 1–11. doi: NODOI PMID:NOPMID

[52] Tóth, A. H., Kelemen, K., Piskóti, M., and Simay, A. E. (2012). *Mobile Phones and Sustainable Consumption in China: An Empirical Study among Young Chinese Citizens. In: China – EU Cooperation for a Sustainable Economy. Corvinus University of Budapest, Budapest, 263–272.* Available at: http://korny.uni-corvinus.hu/cneucoop_fullpapers/s1/agneshofm2.pdf [accessed at: May 21, 2014].

[53] Uddin, A. S. M. I., and Ferdous, J. (2010). Radiation exposure of cell phones and its impact on human health – a case study in South Asia (Bangladesh) and some recommendations. *J. Theor. Appl. Inform. Technol.* 1, 95–97. doi: NODOI PMID:NOPMID

[54] Vergara, E. J., Tehrani, S. N., and Prihodko, M. (2014). *Energy-Box: Disclosing the Wireless Transmission Energy Cost for Mobile Devices. Sustainable Computing: Informatics and Systems.* Available at:

http://www.ida.liu.se/labs/rtslab/publications/2014/VergaraNadjmTehra
ni-EnergyBox.pdf. [accessed at: April 4, 2014].
[55] Vodafone. (2013). *Vodafone Group Plc Sustainability report 2012/
2013.* Available at: http://www.vodafone.com/content/dam/sustainability/
pdfs/vodafone_sustainability_report_2012_13.pdf [accessed at: January
14, 2014].
[56] World Commission on Environment and Development (Wced). (1987).
*Report of the World Commission on Environment and Development:
Our Common Future.* Available at: http://www.un-documents.net/wced-
ocf.htm. [accessed at: January 10, 2014].
[57] World Health Organization (Who). (2013). *What are the health risks
associated with mobile phones and their base stations?* Available at:
http://www.who.int/features/qa/30/en/ [accessed on January 10, 2014].
[58] Yu, J., Williams, E., and Ju, M. (2010). Analysis of material and energy
consumption of mobile phones in China. *Energy Policy* 38, 4135–4141.
doi: 10.1016/j.enpol.2010.03.041 PMID:NOPMID
[59] Zadok, G., and Riikka, P. (2010). *The Green Switch: Designing for Sustainability in Mobile Computing. Paper presented at: The First USENIX
conference on Sustainable information technology (SustainIT'10);
February 22, 2010; USENIX Association Berkeley, CA, USA.* Available
at: http://static.usenix.org/event/sustainit10/tech/full_papers/zadok.pdf.
[accessed at: January 12, 2014].

Biography

M. S. Velmurugan is a Research Scholar of information Technology and
Management at the D.B. Jain College, University of Madras, India. His current
research interests are on e-commerce, e-business, mobile phone, sustainable
development and information technology related topics.